The Social Life of Biogas

Meredian Alam

The Social Life of Biogas

Biogas Users and Their Innovations in Indonesia

Springer

Meredian Alam
Faculty of Arts and Social Sciences
Universiti Brunei Darussalam
Bandar Seri Begawan, Brunei Darussalam

ISBN 978-981-96-3299-2 ISBN 978-981-96-3300-5 (eBook)
https://doi.org/10.1007/978-981-96-3300-5

This Springer imprint is published by the registered company Springer Nature Singapore Pte Ltd.
The registered company address is: 152 Beach Road, #21-01/04 Gateway East, Singapore 189721, Singapore

Preface

I was initially concerned about biogas during fieldwork in a village complex in Yogyakarta, Indonesia in 2009. This investigation on biogas was subsequently initiated. At the time, there were numerous biogas installations or reactors that were situated in close proximity to each other in 11 household units. Each of them was positioned in the courtyard. The 12-month initiative was funded by a university and a foreign non-governmental organization. The two institutions "succeeded" in constructing 11 biogas installations within the predetermined time frame, but not a single household utilized them. However, they continued to be hesitant to handle it, rather than cooking. Upon uncovering this information, I extended my stay at the location from one week to six months in order to investigate the fundamental reasons for the biogas proprietors' refusal to utilize them.

It was discovered during the fieldwork that the refusal of local households is rooted in a number of intriguing facts. First, residents were not contacted regarding the site of the biogas installation, and the location chosen at the time was solely at the discretion of the field manager. Secondly, the assessment and survey did not take into account the number of family members or the amount of energy consumed each day. Thirdly, the assessment procedure did not incorporate cultural values, such as local leadership (village chief, chair of household units, religious leaders, women association leaders), prior to the initial biogas constructions. Fourthly, locals were not invited to or involved in the planning, construction, and training processes. The absence of local residents' participation resulted in a conflict between the biogas owner and the local residents, as the biogas owner was perceived to have received a donation from an external party. This case illustrates a critical component of socio-technology studies: the importance of listening to "cultural voices" that influence the adoption of technology within the community. Finally, after my fieldwork, the situation made me realize that "social acceptance" of biogas technology warrants more investigations.

Social acceptance encompasses far more than just the technical components of biogas. In order for biogas to be accepted and used sustainably, the social practices, culture, and traditions of the society or wider community should be taken into account. At a more micro-level, individual consumption patterns, social setting, jobs, kinship, family relations, financial structure of the respective family need to be included in the consideration and analysis to see the sustainability of biogas technology. And the mission of this book is to understand how biogas technology is innovated, transformed, and adapted into the daily lives of the owners. The writing took place in three countries: Norway, Indonesia, and Brunei Darussalam. This biogas writing and research journey has involved many people over the years, since 2009. I would like to thank Professor Harold Wilhite (Center for Development and the Environment (SUM), University of Oslo Norway) for helping in the design of the fieldwork (he taught me a lot about incorporating elements of "social practices" in the perspective of sociologist Pierre Bourdieu in particular) and Dr. Ingrid Nesheim who always reminded the author to place "cultural voices" as a key element in the ethnographic research. In Indonesia, Professor Suharko (Department of Sociology, Gadjah Mada University) provided an academic support, especially regarding the domestication of technology theory and the role of social institutions in the globalized renewable energy. The finalization stages of this book were mostly carried out in Brunei Darussalam, and for that the author is grateful to have the social support that provides strength in expressing ideas. For this reason, the author is grateful to the Department of Sociology and Anthropology Faculty of Arts and Social Sciences, Universiti Brunei Darussalam (where the author works, learns, and explores knowledge), especially to: Magne Knudsen, Asmali Sulaiman, Shariza Wahyuna Shahrin, Muhammad Arafat Mohamad, and Daniel Wee. From Springer, Alexandra Campbell and Bharath Kumar Dhamodaran have kindly helped the development of this book from the proposal stage to completion one. They responded to my enquiries promptly and ensured everything was on track.

The writing process of this book was engaging, in the sense that my wife, son, and daughter helped me in "translating" transforming every single concept in practical way. They also allowed me to be more grounded and relaxed while writing by allocating more times to play indoor with them. Afterall, I was successfully domesticated.

As a result of contemplation and a catharsis of theoretical study, there are still many elements that need to be examined, particularly the advancement of biogas technology. Consequently, the researcher anticipates that additional research or publications will be conducted on biogas domestication and the innovation of its consumers in a variety of countries.

Bandar Seri Begawan, Brunei Darussalam
September 2024

Meredian Alam

Contents

About the Author

Meredian Alam is an Assistant Professor of Sociology at the Department of Sociology and Anthropology, Faculty of Arts and Social Sciences, Universiti Brunei Darussalam, Brunei. He holds a PhD in Sociology and Anthropology from the University of Newcastle, Australia, an M.Phil. in Culture, Environment, and Sustainability from the University of Oslo, Norway, an M.Sc. in Mental Health from Anglia Ruskin University, United Kingdom, an M.A. in Development Sociology from Gadjah Mada University, Indonesia, and a B.Soc.Sc. in Sociology from Gadjah Mada University, Indonesia. During his 2-year studies at the University of Oslo from 2010 throughout 2012, he had been awarded a research scholarship at the Nordic Institute for Asian Studies (NIAS) at the University of Copenhagen in Denmark to do research on the comparative development of biogas and renewable energy in Nepal, China, India, and Pakistan.

List of Figures

List of Tables

Chapter 1
Prelude

Abstract This introductory chapter argues that community-based technologies, such as biogas, are often promoted and implemented with a technocratic approach that overlooks important social values, traditions, and community dynamics. The success of biogas projects is typically viewed through the lens of technologists, planners, governments, and engineers, who are considered essential to completing the entire biogas construction cycle. In this context, the concept of *governmentality* is employed as an analytical framework to critically examine the role of these actors in the development of biogas technology. This perspective highlights the influence of governmental authority, alongside technocratic players, in shaping collective behaviors, attitudes, and ways of thinking, guiding the community to align with the government's expectations or objectives. While the government utilizes the issue of climate change mitigation to motivate regional and local authorities to boost biogas production, ethnographic data and fieldwork reveal that community social values and long-standing traditions continue to impact decision-making and the sustainability of biogas technology.

Keywords Biogas · Governmentality · Social values · Perspective

1.1 Unveiling Biogas as a "Development" Project

In the widespread implementation of technology, including biogas, decisions are often made solely from the perspectives of technologists, planners, commissioners, governments, and engineers. According to Winkler et al. (2017), *governmentality* serves as an analytical concept for critically examining the role of technology within communities. The governmentality perspective seeks to understand how governments exert influence on communities and societies, as well as how they interact with various forces of power in the modern era. This influence allows governments to shape, reinforce, or even transform collective behaviors, attitudes, and thought processes to align with their expectations or plans (Roubík & Mazancová, 2019). Government power in this context often manifests as a form of disciplinary control.

M. Alam, *The Social Life of Biogas*, https://doi.org/10.1007/978-981-96-3300-5_1

Foucault highlighted the essence of this government action as the "introduction of economy into political practices" (Foucault, 1972). Here, the economy acts as a driving incentive, relying on material factors and rationalities that enable the government to measure targeted communities against predetermined parameters (Russell & Frame, 2013). The involvement of other actors, especially non-state actors, is also crucial in the effective practices of governmentality (Sending & Neumann, 2006). As Catungal and Leslie (2009) explain, these actors help shape and introduce global governance functions within the broader circle of power. This influence is evident, for example, in the role of international non-governmental organizations (NGOs) and civil society institutions in promoting the success of government policies.

When considering governmentality power, two key aspects must be noted: (1) how governments construct policies to achieve their goals and how these policies are applied to the subjects being governed; and (2) how these subjects are treated ethically, ensuring they have equal access to knowledge and are informed about changes in government objectives (Khan et al., 2021). Various global policy practices, such as those for climate change and sustainability (Bäckstrand & Lövbrand, 2006; Baştabak & Koçar, 2020; Keskitalo et al., 2012), communication technology (Díaz-Trujillo et al., 2020; Thrän et al., 2020; Weiland, 2010), and policy science-making (Bevir & Gains, 2011; Fassin, 2011; Tikly, 2003), adopt this concept of governmentality. Drawing from theoretical perspectives on governmentality, this paper aims to explore whether the implementation of biogas projects within Indonesia's climate change regime is ethical. As Christensen (2024) suggests, the focus on ethical concerns is not about categorizing these technologies as inherently good or bad. Rather, it is about how users are positioned as subjects to these technologies often viewed as inferior and whether they have been granted access to essential resources, such as training, technical expertise, and the social knowledge needed to sustain these technologies, along with any policy changes.

Climate change is a global phenomenon, observable, and traceable only through scientific evidence. Its vast and wide-reaching impacts necessitate international and regional collaboration, involving diverse epistemic communities like governments, civil society, and the private sector (Dryzek, 2012; Seyfang & Haxeltine, 2012). In this context, scientifically valid and objective studies are critical for measuring risk reduction (Carley, 2011). However, such practices often overlook humanism by failing to recognize people as central users (Gil-Garcia et al., 2016). Through the lens of governmentality, this paper examines the biogas project in Indonesia to analyze its ethical conduct within the broader climate change regime. Ethics here are not just about binary distinctions of right or wrong but also about addressing the neglect of subjects or grassroots actors involved (Leiter, 2020). Governmentality shows how climate change regimes in Indonesia function to meet mitigation goals aligned with global guidelines, partnering organizations, and Indonesia's own climate change mitigation and adaptation plans. These goals are pursued using neoliberal tools such as numerical achievement metrics, rigid calculations, cost–benefit analyses, and technical approaches to technological governance (He, 2023).

As outlined in Table 1.1, there are three core tenets of power in governmentality. The first is advanced liberal government. The technologies associated with this model involve the use of market principles. For example, the items used in the project were sourced through the marketplace, and their prices are subject to market fluctuations. In other words, self-made materials are excluded due to their unauthorized quality. Secondly, for maintaining project performance, performance technologies were applied, such as key performance indicators (KPIs), benchmarking, audits, and budgeting (Villadsen & Lundberg, 2023). These measurable objectives were implemented as technologies of agency. The second tenet, liberalism, deals with trading and selling mechanisms in the market, market incentives, modern contracts, and the involvement of multiple relevant actors (Pennington, 2023). Since this perspective is grounded in economic rationality, cost, and benefit considerations are central. If these are not met, the project's output is deemed unsuccessful. Market incentives, in this context, refer to monetary rewards for actors who meet specific targets as outlined in the KPIs. The modern contracts involve formal agreements among collaborating partners, providing documentation for the partnership. Liberalism is somewhat inclusive, allowing various actors in the marketplace to participate. The third tenet involves the power to discipline target communities. Rules and regulations are embedded in the implementation process, where community members become subject to surveillance, making them key targets for monitoring and evaluation throughout the project's duration.

The implementation of the biogas program in Indonesia takes place within the broader context of the country's climate change strategy, influenced by international pressures. Indonesia, with its vast forests and "green" resources, as well as its significant oil and gas production, is under scrutiny (Silaen et al., 2020). Biogas is widely recognized for its potential benefits: improving women's health by reducing kitchen smoke, conserving the ecological livelihood of local communities through organic residues, reducing methane emissions that contribute to climate change, and promoting a cleaner environment (Gingerich & Mauter, 2018; Laramee et al., 2018; Madlener et al., 2009; Maosheng & Gehua, 2003; Shane & Gheewala, 2017). However, the role of users and their practices in the adoption of biogas technology is often overlooked. If users' ideas and knowledge were acknowledged, the

Table 1.1 Characteristics of Foucauldian perspective in technology

Authority	Technologies
Advanced liberal government	The use of market principles Technologies of performance: key performance indicators, benchmarking, audit methods, budget evaluations Technologies of agencies: measurable objectives
Liberalism	Trading and selling mechanism of the product Market incentives Modern tract and agreement Multiple actors in the marketplace
Disciplinary power	Use of regulations/rules Use of surveillance (monitoring and evaluations)

Adapted from Oels (2005)

success of biogas adoption would be more likely. Successful adoption can be indicated by the proper construction of the biogas plant, continuous gas production, and users' ability to mobilize their resources—whether financial, local materials, or social networks to ensure the smooth operation of the technology (Alam, 2016; Ghimire, 2013; Henriques & Schnorr, 2010; Widodo et al., 2009). Additionally, key factors supporting successful adoption arise from interactions between users, designers or constructors, and the technology itself (Kivimaa, 2014).

1.2 Overview of the Chapters

According to Jabeen et al. (2020), the co-production of knowledge cannot be separated from the social context in which biogas technology is introduced. This study investigates factors influencing the targeted households in adopting biogas and to what extent they integrate this technology into their daily practices to meet their needs. This book examines how households make decisions regarding biogas technology and identifies key factors that motivate them to adopt it. Additionally, it explores the diverse experiences of users in financing biogas systems and managing their technical operations. While there are no significant regional barriers impeding the adoption of biogas, it remains essential to understand how users perceive and address technical issues in household biogas systems, particularly in light of their prior experiences, local knowledge, established social practices, and livelihood contexts. Furthermore, it is important to explore how users socially construct the technology to reveal the deeper social meanings associated with its use (Arthur et al., 2020; Budzianowski & Postawa, 2017). This book also examines strategies that household users can employ to ensure the continuous operation of biogas systems in their homes.

The Chapter 2 introduces the theoretical perspectives related to consumption, development, and the fundamental principles and implementation of biogas technology in Indonesia. The Chapter 3 reveals the central concept of the domestication of technology. The author's fieldwork experiences, including details of the study site, are discussed in the Chapter 4. The subsequent chapters cover technical and social innovations related to biogas: the appropriation and objectification of biogas in the Chapter 5, and the incorporation and conversion of biogas in the Chapter 6. The Chapter 7 provides the book's conclusion.

References

Alam, M. (2016). Taming the renewables: Actors' innovation in improving the utilisation of biogas for everyday use in agricultural setting. *International Journal of Renewable Energy Development (IJRED), 5*(1). https://doi.org/10.14710/ijred.5.1.57-64

Arthur, R., Baidoo, M. F., Osei, G., Boamah, L., & Kwofie, S. (2020). Evaluation of potential feedstocks for sustainable biogas production in Ghana: Quantification, energy generation, and CO_2 abatement. *Cogent Environmental Science, 6*(1). https://doi.org/10.1080/23311843.2020.1868162

Bäckstrand, K., & Lövbrand, E. (2006). Planting trees to mitigate climate change: Contested discourses of ecological modernization, green governmentality and civic environmentalism. *Global Environmental Politics, 6*(1), 50–75. https://doi.org/10.1162/glep.2006.6.1.50

Baştabak, B., & Koçar, G. (2020). A review of the biogas digestate in agricultural framework. *Journal of Material Cycles and Waste Management, 22.* https://doi.org/10.1007/s10163-020-01056-9

Bevir, M., & Gains, F. (2011). Ideas into policy: Governance and governmentality. *Policy & Politics, 39*(4), 451–456. https://doi.org/10.1332/030557310x550132

Budzianowski, W. M., & Postawa, K. (2017). Renewable energy from biogas with reduced carbon dioxide footprint: Implications of applying different plant configurations and operating pressures. *Renewable and Sustainable Energy Reviews, 68*, 852–868. https://doi.org/10.1016/j.rser.2016.05.076

Carley, S. (2011). The era of state energy policy innovation: A review of policy instruments. *Review of Policy Research, 28*(3), 265–294. https://doi.org/10.1111/j.1541-1338.2011.00495.x

Catungal, J. P., & Leslie, D. (2009). Placing power in the creative city: Governmentalities and subjectivities in Liberty Village, Toronto. *Environment and Planning A: Economy and Space, 41*(11), 2576–2594. https://doi.org/10.1068/a41245

Christensen, G. (2024). Three concepts of power: Foucault, Bourdieu, and Habermas. *Power and Education, 16*(2). https://doi.org/10.1177/17577438231187129

Díaz-Trujillo, L. A., Fuentes-Cortés, L. F., & Nápoles-Rivera, F. (2020). Economic and environmental optimization for a biogas supply chain: A CVaR approach applied to uncertainty of biomass and biogas demand. *Computers & Chemical Engineering, 141*, 107018. https://doi.org/10.1016/j.compchemeng.2020.107018

Dryzek, J. S. (2012). Global civil society: The progress of Post-Westphalian politics. *Annual Review of Political Science, 15*(1), 101–119. https://doi.org/10.1146/annurev-polisci-042010-164946

Fassin, D. (2011). Policing borders, producing boundaries: The governmentality of immigration in dark times. *Annual Review of Anthropology, 40*(1), 213–226. https://doi.org/10.1146/annurev-anthro-081309-145847

Foucault, M. (1972). *The archaeology of knowledge* (Translated from the French by A. M. Sheridan Smith). Pantheon Books.

Ghimire, P. C. (2013). SNV supported domestic biogas programmes in Asia and Africa. *Renewable Energy, 49*, 90–94. https://doi.org/10.1016/j.renene.2012.01.058

Gil-Garcia, J. R., Zhang, J., & Puron-Cid, G. (2016). Conceptualizing smartness in government: An integrative and multi-dimensional view. *Government Information Quarterly, 33*(3), 524–534. https://doi.org/10.1016/j.giq.2016.03.002

Gingerich, D. B., & Mauter, M. S. (2018). Air emission reduction benefits of biogas electricity generation at municipal wastewater treatment plants. *Environmental Science & Technology, 52*(3), 1633–1643. https://doi.org/10.1021/acs.est.7b04649

He, L. (2023). A sociological (re)construction of the reflexive self: A convergence between Foucault and classical Confucian ethics. *Asia Pacific Viewpoint, 64*(2). https://doi.org/10.1111/apv.12370

Henriques, J. J., & Schnorr, W. G. (2010). *Sustainability assessment and implementation of a biogas digester system in western Kenya.* In 2010 IEEE Green Technologies Conference. IEEE. https://doi.org/10.1109/green.2010.5453784

Jabeen, G., Yan, Q., Ahmad, M., Fatima, N., Jabeen, M., Li, H., & Qamar, S. (2020). Household-based critical influence factors of biogas generation technology utilization: A case of Punjab province of Pakistan. *Renewable Energy, 154*, 650–660. https://doi.org/10.1016/j.renene.2020.03.049

Keskitalo, E. C. H., Westerhoff, L., & Juhola, S. (2012). Agenda-setting on the environment: The development of climate change adaptation as an issue in European states. *Environmental Policy and Governance, 22*(6), 381–394. https://doi.org/10.1002/eet.1579

Khan, A., Qyyum, M. A., Saulat, H., Ahmad, R., Peng, X., & Lee, M. (2021). Metal–organic frameworks for biogas upgrading: Recent advancements, challenges, and future recommendations. *Applied Materials Today, 22*, 100925. https://doi.org/10.1016/j.apmt.2020.100925

Kivimaa, P. (2014). Government-affiliated intermediary organisations as actors in system-level transitions. *Research Policy, 43*(8), 1370–1380. https://doi.org/10.1016/j.respol.2014.02.007

Laramee, J., Tilmans, S., & Davis, J. (2018). Costs and benefits of biogas recovery from communal anaerobic digesters treating domestic wastewater: Evidence from peri-urban Zambia. *Journal of Environmental Management, 210*, 23–35. https://doi.org/10.1016/j.jenvman.2017.12.064

Leiter, B. (2020). Foucault as a kind of realist: Genealogical critique and the debunking of the human sciences. *Inquiry, 67*(7), 1–18. https://doi.org/10.1080/0020174x.2020.1712226

Madlener, R., Antunes, C. H., & Dias, L. C. (2009). Assessing the performance of biogas plants with multi-criteria and data envelopment analysis. *European Journal of Operational Research, 197*(3), 1084–1094. https://doi.org/10.1016/j.ejor.2007.12.051

Maosheng, D., & Gehua, W. (2003). Greenhouse gas mitigation benefits of biogas project in livestock farms. *Acta Energiae Solaris Sinica, 24*(3), 386–389. https://europepmc.org/article/cba/559960

Oels, A. (2005). Rendering climate change governable: From biopower to advanced liberal government? *Journal of Environmental Policy & Planning, 7*(3), 185–207. https://doi.org/10.1080/15239080500339661

Pennington, M. (2023). Foucault and Hayek on public health and the road to serfdom. *Public Choice, 195*(1), 125–143. https://doi.org/10.1007/s11127-021-00926-6

Roubík, H., & Mazancová, J. (2019). Small-scale biogas plants in central Vietnam and biogas appliances with a focus on a flue gas analysis of biogas cook stoves. *Renewable Energy, 131*, 1138–1145. https://doi.org/10.1016/j.renene.2018.08.054

Russell, S., & Frame, B. (2013). Technologies for sustainability: A governmentality perspective. *International Journal of Sustainable Development, 16*(1/2), 91. https://doi.org/10.1504/ijsd.2013.053792

Sending, O. J., & Neumann, I. B. (2006). Governance to governmentality: Analyzing NGOs, states, and power. *International Studies Quarterly, 50*(3), 651–672. https://doi.org/10.1111/j.1468-2478.2006.00418.x

Seyfang, G., & Haxeltine, A. (2012). Growing grassroots innovations: Exploring the role of community-based initiatives in governing sustainable energy transitions. *Environment and Planning C: Government and Policy, 30*(3), 381–400. https://doi.org/10.1068/c10222

Shane, A., & Gheewala, S. H. (2017). Missed environmental benefits of biogas production in Zambia. *Journal of Cleaner Production, 142*, 1200–1209. https://doi.org/10.1016/j.jclepro.2016.07.060

Silaen, M., Taylor, R., Bößner, S., Anger-Kraavi, A., Chewpreecha, U., Badinotti, A., & Takama, T. (2020). Lessons from Bali for small-scale biogas development in Indonesia. *Environmental Innovation and Societal Transitions, 35*, 445–459. https://doi.org/10.1016/j.eist.2019.09.003

Thrän, D., Schaubach, K., Majer, S., & Horschig, T. (2020). Governance of sustainability in the German biogas sector—Adaptive management of the Renewable Energy Act between agriculture and the energy sector. *Energy, Sustainability and Society, 10*(1). https://doi.org/10.1186/s13705-019-0227-y

Tikly, L. (2003). Governmentality and the study of education policy in South Africa. *Journal of Education Policy, 18*(2), 161–174. https://doi.org/10.1080/0268093022000043074

Villadsen, K., & Lundberg, J. (2023). Guest editors' introduction to special issue: "Foucault, corporate social responsibility, and corporate sustainability." *Sustainability, 15*(6), 5110. https://doi.org/10.3390/su15065110

Weiland, P. (2010). Biogas production: Current state and perspectives. *Applied Microbiology and Biotechnology, 85*(4), 849–860. https://doi.org/10.1007/s00253-009-2246-7

Widodo, T. W., Asari, A., Ana, N., & Elita, R. (2009). Design and development of biogas reactor for farmer group scale. *Indonesian Journal of Agriculture, 2*(2), 121–128.

Winkler, B., Lemke, S., Ritter, J., & Lewandowski, I. (2017). Integrated assessment of renewable energy potential: Approach and application in rural South Africa. *Environmental Innovation and Societal Transitions, 24*(2), 17–31. https://doi.org/10.1016/j.eist.2016.10.002

Chapter 2
Toward Household Biogas Technology

Abstract Biogas technology in households encompasses both cultural and technical realities that require thoughtful perspectives for understanding its sustainable appropriation. However, socio-technical theories that explain the acceptance of biogas at local, regional, and national levels have received limited attention. To address this gap, this chapter argues that collective norms, cultural practices, the roles of social institutions, shared expectations, and the broader institutional context—such as climate change—are equally instrumental in the development of biogas technology. Here, the sustainability of biogas technology is conceptualized as a co-production process between cultural dynamics and technological knowledge. In a broader context, there exists a climate change mitigation strategy that facilitates the widespread adoption of biogas at regional and local levels.

Keywords Biogas · Collective norms · Cultural practices · Social institutional roles · Sustainability

2.1 The Emergence of Biogas in Indonesia

In modern development, achieving equitable resource allocation is crucial to providing residents of the Global South with access to essential resources (Apergis et al., 2022). To address this, nations are encouraged to adopt the Sustainable Development Goals (SDGs), ensuring that all individuals have access to basic necessities like electricity, energy, and food. However, a growing concern is the unequal distribution of energy access, particularly among those facing extreme poverty in the Global South (Pandyaswargo et al., 2022). Despite significant oil and gas production, and the discovery of new oil resources since 1990, rural populations continue to struggle. Recent studies show that between 70 and 90 million people in these areas still lack the energy required for household power, cooking, and supporting local economies. In light of this, decentralizing energy sources has become essential in addressing this ongoing crisis.

M. Alam, *The Social Life of Biogas*, https://doi.org/10.1007/978-981-96-3300-5_2

One viable approach to decentralizing energy supply is to harness abundant local resources such as animal manure, organic plants for bioethanol, and solar energy (Relva et al., 2021). This book explores the "social acceptance" of biogas, focusing on how methane gas from cattle waste can be converted into alternative energy for domestic cooking. In line with this, the Indonesian government has prioritized optimizing livestock-based energy sources like biogas and fostering community involvement, as outlined in the 2005 Energy Blueprint (Hartono & Maharani, 2021). According to this plan, if local populations benefit from these alternative energy sources, Indonesia's utilization rate could rise from 4.3% to 17% by 2025, bringing the nation closer to its goal of becoming energy-independent in Asia.

Supporting this initiative, the Netherlands-based NGO HIVOS, in partnership with Indonesia's Ministry of Energy and Minerals, launched a biogas program in 2009 aimed at increasing access to biogas technology for rural communities. By 2012, the program planned to build 8,000 biogas installations across eight provinces: Yogyakarta, West Java, East Java, Central Java, Nusa Tenggara Timur, Nusa Tenggara Barat, Bali, and Sulawesi (Budiman, 2021). Local cooperatives and non-governmental organizations (NGOs) are selected by SNV to collaborate with and engage potential customers in marketing operations. These entities are collectively referred to as construction partner organizations (CPOs). The successful implementation of biogas systems in households depends largely on the socio-technical interactions between users and those involved in the construction phase. Additionally, the seamless integration of biogas into daily routines is influenced by the socio-cultural dynamics of the households.

Many discussions highlight the household as a space where technology is adapted to meet daily needs. Technologies such as televisions, cookstoves, and rice cookers are incorporated into homes to fulfill these requirements. To become proficient with this equipment, household members often need to "tame" it—learning how to use it effectively and adjusting it to fit their routines. In this context, Livingstone (2003) and Wilhite (2005) argue that the domestication of technology involves a process of knowledge acquisition, cultural practices, social norms, and values. Carrier and Miller (1999) emphasize the importance of studying the domestication of technology within households, noting that "households play critical roles in shaping the dynamics of energy use in society" (p. 24). This agentive approach positions households as decisive actors in the adoption and adaptation of new technologies.

Against this historical and cultural backdrop, this book explores the impact of biogas on people's daily lives and overall quality of life. Since technology cannot be separated from the socio-cultural elements of society, its presence in homes is influenced by the socio-technical interactions between the technology itself and its users (Clemmensen, 2021). Winther and Wilhite (2014) argue that this raises significant questions regarding the adoption of home energy technologies, the adjustment of values, and why these concepts are so important to individual households (Herrmann et al., 2021). From this perspective, the study examines the social history of biogas technology—from its introduction into homes to how it becomes woven into the fabric of society.

The goal of this study is to highlight and respect users' participation in technological development and their ability to integrate technology into their homes. It also aims to set a precedent for future engineering projects that prioritize transferring technical knowledge to users. The broader objective of this study is to challenge development approaches that overlook the views of technology users, which are critical to the sustainability of the technology itself. This approach frames the relationship between biogas technology, consumption, and development.

2.2 Toward Sociology of Technology and Energy Consumption and Development Energy Consumption

The introduction of biogas technology into the home offers an epistemological advantage by fostering a deeper understanding of the interaction between people and cooking technology within the family space (Lemma et al., 2021; Rahman et al. 2021; Talevi et al., 2022). Traditionally, technological adoption has been viewed through a linear lens—most notably in the adoption theory proposed by Everett M. Rogers. This theory emphasizes that the adoption process depends on knowledge provided by external agents, information flow, centralized implementation, and expert control over the technology (Pansera & Fressoli, 2021; Thompson & Laaser, 2021; Sharma, 2022). However, this perspective overlooks the critical role of users in shaping and influencing technology.

Cultural practices, norms, and even the context of the users have for years been under investigated but in the same vein it provides chance for sociology to work more on the issue mentioned as maintained by British consumption sociologist Elizabeth Shove. Shove (2017) maintains that cultural practices, the analysis of collective norms existing in the societal members, and social expectations are crucial and need to be incorporated in the consumption studies.

Although the deployment of technology often involves technologists or specialists, it is crucial to equally recognize the role of users in fostering creativity in energy use. As Shove et al. (1998) assert, "the structural perspectives inquiry about the social institutional layering of technological adoption and energy choices is critical issues to examine" (p. 296). Biogas, as an alternative technology, may pass through various lifecycle phases, including innovation, the reorganization of practices, and ongoing interaction between users and the technology. Campbell (1995) argues that sociology must study how technology develops within society, emphasizing the importance of cultural context in shaping how people use technology. In particular, Campbell highlights that household consumption is inseparably linked to technology, calling for sociological studies to shift away from a "production-dominated paradigm" toward a more consumption-oriented approach (p. 96).

The present study conducts a qualitative investigation into the use of technical artifacts in rural households, emphasizing the users' perspectives as active agents responsible for modifying, altering, and sometimes challenging existing practices to adopt biogas technology. As a technological artifact, biogas will inevitably undergo modifications driven by user creativity, making it more suitable to their tastes and preferences. This book explores the technical modifications made to biogas technology by users as expressions of ingenuity. Drawing on the works of Callon (1999), Bijker and Law (1992), and Callon and Latour (1992), it highlights the various innovations that arise as technological artifacts settle into daily life. Ultimately, this book argues that collaboration between users and technology is essential for the sustainability of technological systems.

2.3 Addressing "Development"

The implementation of biogas technology presents a distinctive feature in terms of technological acceptance: users are given the freedom to choose reactor types based on their household needs. This demonstrates a significant level of user control over both the technology itself and their understanding of it. Notably, this book introduces innovative examples of alternative technological evolution, moving away from the traditional modernization strategy. In the modernization model, technical professionals often assert dominance from design through execution, treating users as passive participants or mere objects in the process. The concept of modernization is rooted in the belief that development is the transformation of traditional cultures into modern ones, characterized by advanced technology and material wealth (Moore, 1963). From this perspective, the development paradigm tends to view traditions and cultures as obstacles to be controlled rather than as elements to be valued and treated equally.

The ideology behind certain development models, which position themselves as agents of change and saviors of "chosen" nations, has profoundly impacted technological progress, particularly in poorer countries. The core concept of these models often involves undermining the authority of local communities or states, paving the way for exploitation once they have been subjugated. Historically, specialists and developmentalists applying technology in the Global South have operated under the belief that these societies were inherently fragile. As a result, they monopolized the narrative of development, asserting that they alone possessed the knowledge to guide progress. This is often framed as scientific discourse, grounded in Western science, which is portrayed as objective truth (Crewe & Axelby, 2013). However, the application of technology in the Global South is paradoxical and counterproductive to local cultures, as Western knowledge has deeply influenced the thoughts and approaches of techno-scholars in developing nations.

The advancement of developmentalist ideology, as discussed above, continues to be driven by technical application, despite evidence of its partial failures. For example, prior to this research, over 200 biogas installations (or digesters) were made available to the local community through a development program by the Ministry of Energy and Minerals in 2009. However, some of these installations were constructed without homeowners' approval, occupying residential space. Due to geographical challenges, many of these installations became obsolete, and home users eventually stopped using them. No technical repairs were organized to address these issues, and, even worse, local users were disengaged throughout the development process, as they were not properly informed or consulted.

This case highlights a critical flaw: the urgency, benefits, and technical aspects of the technology were not adequately analyzed or communicated to the potential local users. Inadequate evaluation of the technology, based on users' perceived needs and their technical expertise, resulted in these installations being abandoned. This underscores the problematic superiority of specialists in development processes, which has made local communities more vulnerable to the unintended consequences of technology. The nature of underdevelopment and its solutions are often defined by "world-ordering knowledge" imposed by outside experts (Bicker et al., 2003).

Through this study, it is hoped that users' participation in technological development and their ingenuity in adapting technologies within their homes will be respected and appreciated, setting an example for future engineering projects that emphasize transferring technical knowledge to users. The overarching premise of this study is to challenge the development approach that marginalizes the views of local technology users, a shift crucial to the sustainability of the technology itself.

2.4 Biogas Technology, Practice, and Implementation in Indonesia

2.4.1 From Climate Change to Community Biogas

Climate change in Indonesia is primarily driven by anthropogenic effects resulting from industrialization. Over the past decade, the country has responded to climate change through a combination of national policy interventions, international cooperation, and social movements involving community members. These efforts are motivated by the fact that Indonesia is the sixth-largest producer of greenhouse gases globally. As a tropical country with 60% forest coverage, Indonesia was once regarded as the "lungs of the world," but this reputation has diminished over the last ten years. Expansive deforestation and damage to swampland have led to a significant increase in greenhouse gas emissions (Murdiyarso et al., 2019). A major contributor to domestic haze has been the burning of forests in areas designated for the palm oil and paper industries (Tacconi, 2016). As green land and peatlands, which serve as critical carbon storage, diminish, more pollutants are released into the atmosphere rather than absorbed. In addition, the heavy reliance on private cars and motorbikes in Indonesia's megapolitan areas, such as Jakarta, is a leading source of carbon monoxide air pollution (Rachman & Barus, 2019; Thynell, 2018).

Highlighting the track record of the causes and facts surrounding climate change, Indonesia is now under pressure from international bodies and the global environmental movement to reduce its greenhouse gas emissions by 29% by 2030 (Chan et al., 2015). Furthermore, Ozturk (2015) reported that in December 2015, at the UNFCC's Conference of the Parties (COP) in Paris, Indonesia committed to limiting the global temperature rise to 2 °C. To meet this goal, the domestic industrial sector must undertake further sustainable mitigation efforts. From the government's perspective, Indonesia is preparing to reduce exhaust emissions by up to 41% under a "business as usual" (BAU) scenario, contingent on receiving financial assistance and mitigation technology support from developed countries (Alisjahbana & Busch, 2017). Financial constraints remain a significant challenge, as it is estimated that Indonesia requires over $7 billion to meet its climate goals (Kumar, 2016).

Despite a 50% reduction in coal exports, domestic coal consumption has steadily increased. Greenpeace reports that 35% of Indonesia's electricity is currently generated from coal (Kurniawan & Managi, 2018). With a growing population still reliant on fossil fuels, Indonesia must accelerate the adoption of solar panel-based energy to mitigate climate change. Currently, the adoption rate of solar panels remains low, with less than 1% usage in urban areas. In response to this, Indonesia has implemented policies since the administration of President Susilo Bambang Yudhoyono (SBY) to increase the use of solar energy in the country (Anderson et al., 2015).

The Indonesian government has demonstrated its commitment to reducing greenhouse gases. At the 2009 G20 Summit, President Yudhoyono pledged to reduce the country's greenhouse gas emissions by 26% under a Business-As-Usual (BAU) scheme by 2010 through unilateral actions (Enrici & Hubacek, 2016). Indonesia's ambitions have continued to grow, implementing seven key mitigation strategies, including promoting low-emission transportation, developing renewable energy sources, enhancing energy efficiency, initiating carbon sequestration projects in forestry and agriculture, halting deforestation and land degradation, maintaining peatland sustainability, and improving environmental management. However, achieving the expected quality of outcomes in line with the UNFCCC's goals has proven difficult (Ahmed & Wang, 2019). One major challenge has been the polarization and political instability within Indonesia's legislature (Sforna, 2019).

In 2011, a presidential decree was issued to address climate change, leading to the introduction of the Greenhouse Emissions Reduction Policy (Buditama, 2016), which serves as a national framework for implementing Nationally Appropriate Management Activities (Nachmany et al., 2015). Greenhouse gas control measures are applied at both the national and regional levels, with provincial governments instructed to create Gubernatorial Regulations for Provincial Action Plans to mitigate the impacts of climate change (Nachmany et al., 2015). The implementation framework is cross-sectoral, requiring collaboration between ministries to tackle the economic, social, environmental, and political dimensions of the issue. Cooperation spans forestry, industry, agriculture, energy, infrastructure, and economic instruments such as taxation (Wibowo & Giessen, 2015).

Despite ongoing challenges, the government has shown a strong commitment to climate change mitigation. For instance, the establishment of the Directorate General of Climate Change within the Ministry of Forestry and Environment is a significant step forward. In 2015–2016, Indonesia also introduced the Green Short Run State Planning (GSRSP) as part of a new "green economy" framework. This framework aims to mitigate climate risks while promoting inclusive and sustainable growth, improving quality of life, and addressing disaster preparedness. In response to political pressure, Indonesia has embraced multilateral cooperation for greenhouse gas mitigation. BAPPENAS, the national planning agency, became the coordinating body for these efforts, and Biogas Rumah was launched in 2010 under the Clean Development Mechanism (CDM). The CDM scheme allows developed countries to implement greenhouse gas reduction activities in developing countries, such as renewable energy projects and reforestation, in exchange for tradable carbon credits. Due to its significant agricultural and livestock activities, Indonesia was seen as an ideal CDM location. The Indonesian government's openness to international funding and development donors has facilitated climate change mitigation programs, though the country's weak bargaining position has made it more of an object of international initiatives rather than a leader in these programs. The Ministry of Energy and Minerals, in addition to BAPPENAS, plays a vital role in implementing climate change mitigation projects. Under Indonesia's decentralization policy, renewable energy projects are managed at the provincial level, allowing for collaboration with non-governmental organizations and offering provinces more autonomy in their climate action efforts.

2.5 Biogas Technoscape in Indonesia

The unpredictable prices of kerosene and LPG in Indonesia have raised concerns for rural communities, particularly for low-income households. This situation highlights the need for alternative energy sources that can support both economic improvement and environmental sustainability. One promising solution is to encourage communities to adopt biogas for domestic use. A 2008 study by the Netherlands Development Organization (SNV), funded by the Royal Embassy of the Netherlands, explored this possibility, laying the groundwork for further biogas initiatives in Indonesia.

The implementing agency for this program is HIVOS (Humanist Institute for International Cooperation), working in consultation with SNV. Before launching the Household Biogas Program, known as "Biogas Rumah" (BIRU), SNV conducted a feasibility study to ensure the selected locations had ample sources of cattle dung and were geographically suitable for durable technical construction. Based on this study, three provinces were recommended as pilot locations: West Java, Central Java, Yogyakarta, and East Java. In West Java, the project was implemented in Lembang (West Bandung) and Bayongbong (Garut), while Mojosongo (Boyolali) in Central Java and Jabung (Malang) in East Java were also chosen.

The program actively involves non-government organizations and educational institutions as construction partner organizations (CPOs). This collaborative approach allows various stakeholders to engage in the construction process and achieve the program's objectives. Initially, CPOs were tasked with constructing at least 100 biogas units in their selected areas, adhering to predetermined quality standards. Additionally, a Training Partner Organization (TPO) was appointed to conduct construction training for the involved parties. Finally, Lending Partner Organizations, which include regional banks and microfinance institutions, are responsible for providing access to credit for prospective users (Biogas Rumah [BIRU], 2024).

In 2010, the biogas program, known as the Household Biogas scheme, was launched in Indonesia. The initiative was spearheaded by two key players: Yayasan Rumah Energy (Rumah Energy Foundation/YRE) and HIVOS Southeast Asia. Additionally, the Netherlands-based technical consultant SNV (Stichting Nederlandse Vrijwilligers-Netherlands Development Organization) was integrated into the program's framework, serving as a supervisor for the technical sustainability of biogas reactors. These three non-governmental organizations collaborate in an alliance to promote the program. On the government side, the state planning agency, BAPPENAS, is responsible for reporting on the achievements and positive impacts of humanitarian projects at the national level.

Indonesia's commitment to climate change mitigation and adaptation gained momentum in December 2007 when it hosted the COP-13 (Conference of the Parties) in Bali, an event notably supported by President Yudhoyono, who inaugurated the National Climate Change Council. Since then, Indonesia's dedication has strengthened further, as evidenced by its ratification of the Kyoto Protocol, committing to reduce greenhouse gas emissions by three percent by 2030. However, the country faces significant environmental challenges, including high rates of deforestation and forest degradation. Balancing these environmental concerns with the need for economic growth through industrialization and the sustainable use of natural resources, such as forests for pulp and paper production, remains a complex issue. Nevertheless, Indonesia must advance and solidify its commitment to environmental improvement alongside other countries worldwide, fostering both government-to-government collaboration and partnerships between public and private sectors. The Reduction of Deforestation and Forest Degradation (REDD+) initiative, supported by the Norwegian government as a donor in this global environmental effort, has become a focal point for Indonesia (Do & van Noordwijk, 2023). Despite criticisms from global environmentalists regarding the potential disruption of indigenous social structures due to cash incentive provisions, allegations of systematic corruption within regional bureaucracy, and a lack of financial transparency, REDD+ has empowered Indonesia with a politically advantageous position on the international stage (Vijitharan et al., 2024). This has allowed the government to forge stronger international alliances in environmental preservation through various multilateral agreements (Dugasseh & Andersen, 2024). Consequently, the repertoire of pro-environmental policies adopted by the Indonesian government has gained recognition from developed nations.

Indonesia is viewed as a willing participant in international climate change mitigation efforts, and its climate regime is considered robust. However, this approach often presents more drawbacks than benefits for local communities. Climate change, being a science-driven issue, remains largely the domain of scientists, engineers, climatologists, chemists, and economists, who prioritize objectivity and hard scientific methods. This emphasis on objectivity can create a disconnect between these experts and the communities they aim to protect, educate, and involve in building sustainable futures. Such a paradigm can stifle the implementation of technology and climate change innovations, diminishing social benefits for communities and undermining their roles. As a result, society may unconsciously be positioned as inferior, becoming objectified in the process.

Biogas technology is labor-intensive, requiring a high level of skill and technical competence from users. Effective management includes tasks such as filling the reactor with cow dung, monitoring gas volume with a manometer, and periodically cleaning the reactor by removing hardened residual pulp from the outlet. The complexity extends further, as biogas stoves also need regular cleaning, particularly in the fire output section, which is prone to corrosion due to methane contamination from excess sulfur and soft zinc. Unfortunately, this stove design lacks sulfur catchers to mitigate corrosion, indicating that current innovations are not yet fully effective. Moreover, the development of biogas systems has often followed a technocratic, non-user-oriented paradigm. This approach has led to the unilateral reduction of physical materials for the reactor by designated construction material suppliers, compounded by delays in material supply (Williams et al., 2024).

As a technological artifact, biogas systems necessitate that users enhance both their social and technical competencies to ensure sustainable performance and positive impacts on agricultural output quality, welfare improvement through reduced oil consumption, and enhanced health and environmental conditions using by-products (Ni, 2024). Training in the use, maintenance, and repair of biogas systems is crucial in this context. By educating users on these aspects, it is expected that a strong, immersive relationship will develop between them and the technology. If such a relationship materializes, domesticated biogas could seamlessly integrate into people's daily lives (Robinson et al., 2023). However, as highlighted in this section, many users face disenfranchisement due to the technical and social complexities surrounding biogas systems. As previously mentioned, umbrella organizations pursuing carbon credits and compensation often focus on the quantity of installations and their international reputation rather than the quality of the end product.

This renewable energy project specifically targets a particular social group: farmer groups, which serve as the main entry points for local construction partners in each province of Indonesia. Within the framework of the biogas project commissioned by the Indonesian government, these farmers are often reduced to mere objects of repeated failure. Despite receiving a bonus of $1 million per biogas reactor as compensation, this amount does not adequately account for the time and risk sacrifices endured by construction partners, which include locating potential users and managing the logistics of construction materials. Although local construction partners are not pressured to install a specific number of reactors, they are tasked with reaching as many users as possible.

Farmers are part of cohesive tactical groups characterized by strong organic solidarity and social trust among members, making them easier to convince to adopt biogas technology. This unity means that local construction partners can quickly obtain a list of prospective users within just a few days, often with little more than a presentation on the benefits of biogas during a meeting and a discussion with the group leader. If a member has at least 3×3 m of vacant land located 10 m from their house, they readily register as potential users. The availability of cows is also considered, as this allows them to source cow dung from neighbors. The enthusiasm among farmer group members is evident in the high registration rates per week, which is encouraging for the coordinating staff at the ministry's regional office. For these officials, the quantity of installations completed is more critical than the quality of the biogas systems post-installation.

A technological fallacy emerges in the assumption that biogas systems are easy to use and install, leading to inadequate socialization and orientation for farmers. In one instance, an officer with an engineering background—who lacked familiarity with social dynamics and community engagement—failed to recognize that 90% of the installed biogas cooking stoves did not ignite due to various issues, including broken reactors and insufficient cow dung. Moreover, this officer refrained from conducting field checks or monitoring user training. At this stage, it becomes clear that regular interaction with users and their initiatives to maintain and sustain the technical aspects of biogas systems are vital for successful implementation (Izadi et al., 2024).

The post-construction stage also entails formal user training (Budiman & Smits, 2020). This mandatory phase is designed to equip users with essential skills and practical knowledge, including how to contact construction officers for maintenance. However, through observation and rigorous interviews with users, it became evident that they faced significant challenges in accessing spare parts for their biogas stoves. One of the main issues was rusted gas pipelines, which were damaged further by excess water that accompanied the methane. As of October 2011, the Rumah Energy Indonesia Foundation reported that engineering experts were still conducting trial-and-error studies to address these technical glitches. Unfortunately, the trial-and-error period at one Indonesian university has yet to be completed. Releasing products into the market that have not finalized their testing exposes users to potential risks.

Since the initiation of the Blue Biogas Program in late 2010, over 1,000 stoves have been distributed and adopted by household users. However, alarmingly, only two out of every ten reactors are operational (Budiman et al., 2018). During a meeting in Jakarta attended by representatives from YRE and HIVOS Southeast Asia, this troubling statistic was presented, but there was no comment from the officials. While they acknowledged the issue, they did not address any plans to retrieve the non-functional stoves from users who had relied on them for months (Budiman & Smits, 2020). These technological fallacies reflect the convenors' indecisiveness and a failure to consider the cultural and social dynamics of the villagers. This situation underscores the importance of co-production efforts, where technology developers and users or target communities must collaborate to ensure that technical details are tailored to local contexts (Ali et al., 2022).

The results of these observations and interviews were unveiled in 2012. The author's fieldnote data revealed:

> A majority of household users involved in the project under the Ministry of Mining and Energy, in collaboration with the Rumah Energy Foundation, lacked understanding of technical terms and the necessary procedures for checking stove damage. These technical terms are crucial for identifying defective spare parts. Although posters were displayed on kitchen

> walls detailing methods for filling the reactor with cow dung and water, as well as information about stove and reactor spare parts, users often did not follow these guidelines in practice. Many users admitted that shortly after the installation of the biogas reactor, officers would provide them with books, posters, and their contact numbers but would fail to explain the technical aspects of the reactor in depth. The lack of adherence to the handbooks and posters cannot solely be attributed to the users or the construction partner organizations; rather, it reflects a broader issue of inadequate training and insufficient skill development preparation by YRE and HIVOS Southeast Asia, representing the Dutch government. Construction partner organizations, in this context, serve merely as field executors of the training policies established by these larger organizations. By equipping construction partner organizations (CPOs) with pedagogical competencies, the likelihood of biogas damage could be reduced. Following a meeting in October 2011, YRE, representing HIVOS Southeast Asia, promised to address these technical shortcomings in the coming years.

At this point, as Robinson et al. (2023) studies, the technical details might be interpreted differently by each involved party as their education and experience level in working with such technology is different.

The communication aspect between users and construction partner organization (CPO) officers presents another challenge in the biogas project (Bößner et al., 2019). Ideally, users and CPOs should share a synchronous relationship. CPOs are recruited from the local community with the expectation that they will be trusted and can leverage existing cultural bonds to increase the number of household users. However, while an engineering background is typically a requirement for officer recruitment, many of the individuals the authors interacted with were non-engineers and lacked experience in handling construction projects. Consequently, they primarily acted as promoters rather than fulfilling their dual roles as both promoters and technical contacts.

Furthermore, most household users refrained from complaining to or consulting with the CPO officers when their reactors or cookstoves malfunctioned. The fear of losing social prestige was a significant factor deterring users from reaching out for help. At the same time, users expressed deep disappointment with the complexity of the biogas system, which further discouraged them from seeking assistance. Compounding these issues, it was found that users often changed their mobile phones every three months, resulting in the loss of contact information for the officers or YRE staff. Given that biogas is not a priority for fulfilling their basic needs, this perception further diminished their motivation to stay in touch with their respective CPO officers.

In the framework of top-down climate impact mitigations, the Rumah Energy Foundation's achievement was earning a gold certification. It is a target that is mandatory to be achieved every year. This internationally reputed certification is an objective requirement so that their biogas projects are recognizable and claimed as measures and a part of concerted political efforts to reduce greenhouse gasses. The number of successfully installed biogas is the main parameter and this data is obtained from all Rumah Energy Foundations across Indonesia. However, the quality of the biogas usage and damage in installations are not taken into consideration in estimating GHG (greenhouse gases) reductions. In other words, figures and field data from biogas installations are the final output that are considered and therefore pursued. The quality of practices of user biogas in the situation described aligns with the experiences of Quality Assurance (QA) officers from the Rumah Energy Foundation. According to one officer, the quality of biogas installations has declined since the Ministry of Energy and Mining took over the program. He noted that the biogas scheme under the Rumah Energy Foundation, which allowed community members to voluntarily register as users, produced better results. Frustrated with the current state of the program, the officer expressed his inability to address the quality issues in the field, particularly concerning the products distributed by the government.

Although the regional construction team was aware of these setbacks, they were powerless to implement changes without additional funding, which required approval from the central office in Jakarta. The officer emphasized that their hands were tied, as they were operating within a rigid, technocratic system that prioritized a deterministic approach to project management. Although his background is chemical engineering, which should have equipped him to provide technical support to household biogas users, the QA officer opted to remain a "silent observer." His frustration stemmed from both the deteriorating product quality and the limitations imposed by a system focused more on technical outcomes than on the human impact of the biogas program. Further concerns arose during a consultation between the chief of construction and the writer during an evaluation and monitoring session. It was suggested that by offering incentives and providing access to biogas technology, the program hoped to empower community members to contribute to reducing greenhouse gas emissions, particularly methane. However, the lack of user involvement in training sessions and community meetings, combined with an indifferent response to widespread issues like damaged stoves and biogas reactors, highlighted a deeper disconnect between the stakeholders.

The top-down approach, which included private stakeholders as part of corporate social responsibility initiatives, also seemed ineffective in delivering tangible benefits to the wider community. For instance, a private hotel in Indonesia agreed to collaborate with regional office by purchasing organic fertilizer, a by-product of the biogas system, for its ornamental plants. Yet, this partnership involved only a small number of households—just 1–5 and appeared to serve more as a publicity effort for the biogas program and its international sponsors than as a meaningful community initiative.

This context does not support the objectivity of measurement, as engineering science logic applies here, where user experiences and socio-cultural subjectivity are difficult to quantify using technical parameters. Rather than being viewed as an opportunity, these socio-cultural aspects of household biogas users are often perceived as obstacles. Concurrently, the mechanistic approach to biogas management seems wasteful. Although household users are not required to pay for the installations and are only asked to provide land for the reactor's construction, biogas has become a dormant artifact—a neglected asset, akin to obsolete objects in a museum of antiquity. In many homes, biogas stoves sit unused in kitchens, symbolizing this stagnation.

In this context, gold certification appears to be part of a political narrative within the climate change regime, encroaching on private spaces and assets. As Sealey-Huggins (2017) suggests, "climate change imperialism" aptly describes this criticism of gold certifications. This critique is closely tied to the calculation of greenhouse gas emissions and claims within the Clean Development Mechanism (CDM). For this biogas project to be recognized as part of climate change mitigation, gold certification must be validated by external evaluators, typically experts in climate economics. Gold certification also opens the door for organizations like Rumah Energy Foundation to trade carbon credits internationally. Developed countries can purchase these credits and present them at UNFCCC as part of their climate change mitigation efforts.

Furthermore, ethical concerns arise when household users are not adequately informed about their role in the greenhouse gas reduction scheme. As participants who provide land for reactor construction, these users contribute to the carbon offset market, yet they may be unaware of this fact. It is likely that most household biogas users have never been informed that their biogas installations are counted toward reducing greenhouse gases. While knowledge of methane gas and climate change is often reserved for elites, it could and should be shared with local communities.

Developing creative and effective media outreach strategies to educate household users about climate change is essential for any partner organization involved in biogas projects, including the Rumah Energy Foundation. Although YRE has focused primarily on technical development, additional projects like organic fertilizer training have also been initiated to extend the benefits of these biogas initiatives. Despite the project's intentions, the reality on the ground suggests that it has provided limited benefits to household users, with the blame for its failures often placed on them. From a social class perspective, these users primarily belong to the working class, with most engaged in paddy farming and some working as traders at the local wet market. Field observations indicate that household users are often regarded as passive subjects in the technology's implementation, with their social background seen as a barrier to understanding the technical aspects of biogas (Bresnihan & Brodie, 2024). In essence, the project treats household users as uninformed or ignorant. This dynamic reflects what sociological perspectives describe as social class abuse, highlighting the knowledge inequality inherent in the project.

Moreover, the government's approach remains "business as usual," with success measured by the number of reactors built and physical installations completed, rather than by the actual impact on users' lives. This form of governance reflects an unethical bureaucratic and technocratic mindset, contributing to the perpetuation of technological fallacies in Indonesia's biogas sector. Biogas installations vary across countries, influenced by factors such as geography, economy, and program goals. For example, polyethylene biogas systems, which utilize long-stretched plastic containers, are constructed at a relatively low cost, but their durability is often questionable. In contrast, some users opt for doom-type gas holders, which are made from cement, sand, and other waterproof materials. This type is believed to be more durable and weather-resistant. However, the use of more robust building materials typically results in higher costs and requires intensive maintenance.

This book focuses on the development of household biogas, particularly in rural Indonesia, where doom-type biogas systems are preferred. Given that the operation and construction of biogas systems can be complex and require advanced technical expertise, this section emphasizes the governance and execution of biogas construction in the field, along with its technical operations and practical applications for daily living. Ryan and Vivekananda (1993: 20) define appropriate technology as technology that can be operated using locally obtainable materials within specific rural contexts. This type of technology is significant because it provides households with multiple benefits that contribute to the well-being of local communities. The main principle of appropriate technology is characterized by the use of local materials for construction, particularly in specific rural settings where its routine usage is informed by local experiences and connected to the existing cultures that have long shaped community interactions.

Biogas exemplifies appropriate technology that relies on local materials, commonly used in rural housing construction. It reflects typical technological practices in rural areas, where the materials for construction and daily feeding into the biogas plants are often abundant. Furthermore, Quadir (1995) highlights the compatibility of biogas appliances for rural communities, suggesting that the characteristics of rural areas in many developing countries in Asia and Africa—with their abundant livestock—allow biogas technology to flourish more rapidly than in urban settings. Regarding the endurance and durability of biogas systems for household use, extensive research and development have been conducted. This innovative biogas technology has been tested and implemented in targeted countries, including Vietnam, Ethiopia, Pakistan, Nepal, Cambodia, and Bangladesh, where fixed-dome systems have been installed underground.

2.6 How Household Biogas Technology Works

The supply of local materials used for biogas construction, as well as the quality of cow dung, plays a critical role in ensuring the sustainability of this technology. Further studies that examine the interplay between adoption and local material supply are elaborated upon by Akinbami et al. (2001) and Parawira (2009). Through their investigations and observations of various global biogas practices, these scholars assert that biogas technology is locally viable and does not require sophisticated support or extra equipment for maintenance. Additionally, the skills possessed by local communities are typically well-suited to the demands of this technology. All components of biogas technology, including construction materials, can typically be sourced from nearby towns and local supply chains. This local sourcing helps reduce production costs, allowing for potential savings that can be redirected to finance other essential needs (Henriques & Schnorr, 2010; Purohit & Kandpal, 2007). However, biogas systems require intensive maintenance to ensure their long-term sustainability.

The fixed-dome biogas system can sustain operation for 15–30 years, provided that household users maintain it properly and follow the recommended techniques. To operate the biogas system, users need only to feed it with cow or pig dung mixed with water in a 1:1 ratio (e.g., 10 L of water mixed with 10 kg of cow dung). Therefore, each household should ideally have at least two cattle or cows to ensure a consistent daily supply of feed. From a technical perspective, the biogas digester is designed according to standardized systems and consists of several components: an inlet, a reactor, a gas storage unit, a gas carrier system, and a slurry pit. The inlet serves as a cemented containment area where water and animal dung are mixed before being transported to the biogas reactor. The outlet is a chamber designated for undigested mixtures, while the bio-slurry pit contains solid residue that is free of methane gas and rich in organic nutrients.

The biogas system is user-friendly when the correct procedures are followed. First, users must mix water and dung in equal proportions (1:1) in the inlet. This mixture is then fed into the plant or reactor through a designated underground pipeline. Inside the reactor, methane gas is produced and stored in the upper section of the dome. This gas is then ready to be channeled to the kitchen for cooking through transport pipelines.

The residue from the gasification process is released into an outer chamber known as the bio-slurry pit. It's important to clean this pit when it reaches full capacity to ensure efficient operation. Unlike conventional cooking equipment, biogas infrastructure necessitates distinct locations for constructing the digester and the biogas reactor, along with other essential components such as an intake system for producing sludge and a slurry pit or container for storing organic waste. Regular maintenance is crucial; this includes cleaning the slurry container periodically. The cleaned residue can then be utilized as organic fertilizer, providing an additional benefit to the system. Figure 2.1 is shown.

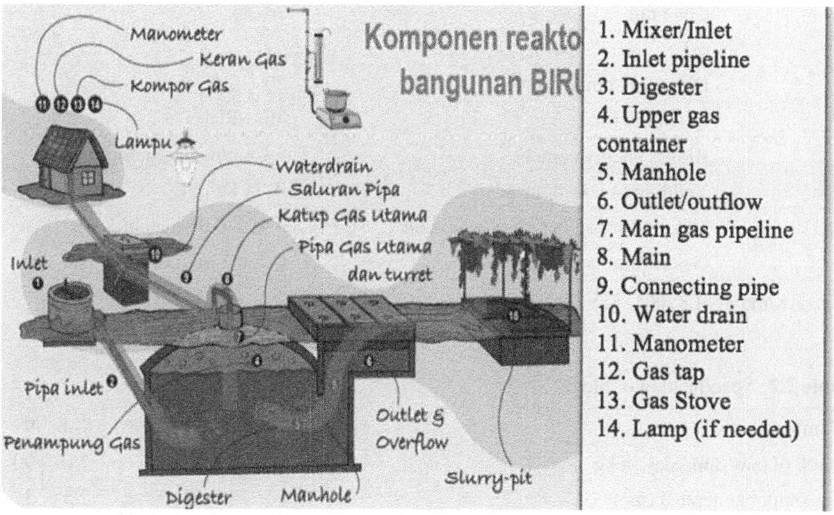

Fig. 2.1 The structure of biogas technology (*Source* http://www.biru.or.id/en/index.php/digester/)

2.7 Design of Biogas Plant

The design of biogas plants must comply with specific technical requirements to ensure sustainability. There are 4 (four) essential parameters to consider are C/N Ratio, pH, Digestion Temperature, and Retention Time. The C/N Ratio refers to the proportion of carbon to nitrogen present in organic matter. Optimal gas production occurs when the C/N ratio is between 20 and 30. The pH measures the acidity or alkalinity of the input material; a pH of 7 is considered neutral, while values above 7 indicate alkalinity. For effective gas output, the pH should ideally be maintained between 6 and 7. Digestion temperature is crucial, as it relates to the optimal conditions for the location of the biogas plant. A temperature range of 20–35 °C is considered best for facilitating gas production. Therefore, biogas technology is not feasible in a cold climate. The retention time is an average time that a given quantity of input remains in the digester (Karki et al. 2005:13). Table 2.1 summarizes the technical requirement:

Each biogas size has different specifications to be considered for its further operations, such as summarized in Table 2.2.

For effective gas output, the pH should ideally be maintained between 6 and 7. Digestion temperature is crucial, as it relates to the optimal conditions for the location

Table 2.1 Design parameters of biogas plant

No	Parameter	Value
1	C/N ratio	20–30
2	pH	6–7
3	Digestion temperature	20 -35 degree Celsius
4	Retention time	40–100 days
5	Biogas energy content	6 kWh/meter cubic
6	Yielding of cow dung	9–15 kg/day
7	Gas production per kg of cow dung	0.023–0.04 m cubic
8	Gas requirement for cooking	0.2–0.3 m3/person

Source Karki et al. (2005: 31)

Table 2.2 Specification of biogas

Number of cow	3	4.5	6	7–8	9
Stock of cow dung/day in kg	30	45	60	75	90
Gas output generated (m^3)	1	1.5	2	2.5	3
Length of cooking time on biogas cookstove	4–6	6	8	10	12
Estimation of total price of construction materials (in million IDR)	5.7	6.3	7	8	8.8
HIVOS subsidy (million IDR)	2	2	2	2	2

Source Interview with AWFFDFM

of the biogas plant. A temperature range of 20–35 °C is considered best for facilitating gas production. The varying sizes of the reactors require users to pay different rates. However, HIVOS provides each user with IDR 2 million, which covers the costs of construction labor, galvanized pipes, biogas cookstoves, a 36-month guarantee, and post-construction training. Users are only responsible for paying the remaining balance.

Access to local materials for constructing biogas technology infrastructure is closely linked to the economic situation of the adopters. In developing countries like Indonesia, where approximately 60% of the population resides in rural areas and depends on agriculture, biogas technology is more likely to be adopted by middle-income farmers. These farmers typically have better access to economic resources and land ownership (Devadas, 2001; Ni & Nyns, 1996). If the supply of materials is deemed sufficient, the construction process, carried out by skilled and well-trained laborers, becomes crucial for both the construction stages of the biogas plant and its ongoing maintenance (Pal, 2005). It is important to evaluate labor costs while also considering a household's financial capability to afford these services (Purohit & Kandpal, 2007). In addressing concerns about employing labor, Walekhwa et al. (2009) suggest that involving family members can help maintain the sustainability of the biogas system. This involvement can also strengthen emotional ties among family members.

Access to materials is a key factor in successfully adopting appropriate technology, as it pertains to the availability of local resources. It is argued that locally available raw materials are more influential and empowering than those sourced externally. Utilizing local resources for the operation of appropriate technology is believed to improve the lives of household users and the wider community. Thus, revitalizing locally-based skills and experiences is essential. Local expertise can help reduce the "assembly cost" of technology, allowing people to use their own resources to expedite installation without the need to hire outside experts.

After returning from fieldwork, the theoretical perspective on appropriate technology becomes prominent in this study, as the findings align with several aspects discussed under this concept. It was also found that the continuity of biogas operation in a household setting relies heavily on sufficient management by the users, which should be a shared responsibility with policymakers. Since household biogas systems are user-oriented technologies, prioritizing sustainability components that empower users to autonomously manage and maintain this technology is essential. This includes user-friendly designs for the biogas systems, technical training for users, reliable suppliers of construction materials and spare parts, and regular monitoring and evaluation of everyday usage. However, in Indonesia, various challenges hinder the consideration of users' experiences and subjectivities in the improvement process (Mulyono et al., 2024).

Previous studies have identified challenges in the troubleshooting of biogas installations at home, providing valuable lessons for biogas engineers, construction agencies, financing organizations, and policymakers. These issues highlight the missing links in engaging users' experiences both prior to and after construction. Firstly, it is pointed out that inadequate maintenance has rendered many biogas plants inoperable, discouraging their use and complicating efforts to market the technology to new customers. Secondly, a study on household biogas in Indonesia revealed that about 60% of installations suffered from poor quality, likely due to farmers not properly mixing or watering the feces. Training programs aimed at enhancing digester efficacy, particularly for women who are often the primary operators, could significantly improve digester management and biogas output. Additionally, the lack of technical training provided to users presents another barrier, illustrating a tendency among policymakers and biogas engineers to view users as mere objects rather than active participants. Fragmentation in Indonesia's biogas projects has also impeded the diffusion of biodigesters and the transfer of information.

Problems with both technology concepts often become apparent a few weeks after implementation in certain areas. The root cause of these issues is the failure to strictly enforce technical building standards. Changes in cooperation and power allocation within Indonesia's biogas governance architecture have led to recurring regime adjustments from 2007 to 2017. These transitions included: administrative fragmentation from 2007 to 2009, conflictive fragmentation from 2010 to 2013, cooperative fragmentation from 2014 to 2016, and a decrease in fragmentation in 2017.

Between 2007 and 2009, many ministries implemented their own biogas initiatives without the involvement of HIVOS. In 2010, HIVOS established the BIRU program, which had successfully installed over 11,000 digesters by 2013. Simultaneously, the Ministry of Energy and Mineral Resources (MEMR) widely promoted the 4K digester. The distribution of biogas funding increased as BIRU attracted additional donors, allowing the MEMR to secure a larger share of the available funds for biogas initiatives.

BIRU's efforts to educate community project officers (CPOs) led to a rise in the number of skilled biogas construction workers, resulting in 16,000 biodigesters now available for use. However, the establishment of the Directorate of Renewable Energy at the MEMR and the creation of BIRU further fragmented existing biogas initiatives. This redistribution of authority accelerated the rollout of programs and increased the frequency of their dissemination. However, a lack of unity among decentralized authorities has led to discordant splintering within the biogas governance framework between 2010 and 2013. In such an architecture, distributed power can only be harmonized through effective coordination and collaboration. Beginning in 2014, collaborative efforts started to take shape, resulting in an increase in biogas initiatives from 2014 to 2016 due to BIRU's partnerships with additional government agencies. The distribution of biodigesters surged from 16,000 in 2013 to 37,000 in 2016 as a result of this collaborative fragmentation.

User education on biogas initiatives also benefited from these partnerships, with a growing emphasis on standardizing training. HIVOS and YRE developed training criteria through the BIRU initiative, educating several community project officers (CPOs) on how to instruct biogas end-users. These CPOs adopted the same training standards when collaborating with other biogas initiatives for ministries and municipal governments. However, in 2017, the government implemented a new centralized strategy, which reduced decentralization in the biogas administration structure. This adjustment led to a decline in the number of biodigesters distributed.

Cow dung plays a crucial role in the biogas technology within villages characterized by agricultural practices and the cultural norms of Javanese society, where mutual assistance is a deeply rooted value. Users' experiences with biogas technology can vary significantly based on these cultural contexts. Ryan and Vivekananda (1993) highlight the interplay between biogas technology and local culture, emphasizing its importance for the sustainability of daily operations and usage. When considering biogas technology as an example of appropriate technology, it is well established that cow dung serves as an essential organic material for its operation. The use of cow dung for biogas production not only sustains the life cycle of nature but also integrates with the sustainability of the global ecological system (Capra, 2003). Furthermore, the relationship between humans and nature is an inseparable aspect of this discussion in appropriate technology (Ryan & Vivekananda, 1993). From this perspective, appropriate technology is interconnected with a holistic worldview of nature, promoting the integration of beliefs, social forms, and harmony with nature alongside the organization of physical materials in accordance with scientific principles (Ryan & Vivekananda, 1993: 52).

2.8 Understanding the Involvement of AWFDFFM in "Biogas BIRU" Program

Organizing the biogas program presented a new challenge for the Association of the World Food Day Farmers and Fishermen's Movement (AWFDFFM), as the organization had never engaged in similar activities before. However, a logical framework emerged because the program aligned with the organization's vision. From an agricultural perspective, the implementation of biogas is consistent with AWFDFFM's ecological values, aimed at revitalizing organic farming. This initiative aims to foster greater independence within the community from kerosene-based fuels, which can trap the local economy in cycles of dependency. This section will explore the historical trajectory of AWFDFFM since its establishment in 1992 and will subsequently discuss its responsibilities as a construction partner organization (CPO) in the implementation of the biogas program in Bantul Regency, Indonesia.

In the implementation of the Household Biogas Program ('Biogas Rumah'), several civil society organizations in Indonesia have played a vital role in promoting and accelerating the growth of biogas systems across the country. These organizations are considered strategic actors, as they possess a deep understanding of local and community contexts, enabling them to effectively engage with and support the needs of the population. AWFDFFM was established in January 1992, with its founding statute, the Ganjuran Declaration, emphasizing that agriculture, marine culture, and rural development must be ecologically sound, economically feasible, and culturally appropriate. These principles are seen as key to sustainable human development. In

1994, groups of Indonesian fishermen voluntarily joined the organization, prompting it to expand its advocacy efforts to include the needs of fishermen.

Concerns regarding environmental degradation have led AWFDFFM to identify several critical challenges, such as dependence on pesticides, fertilizers, seeds, technology, and capital. The use of synthetic fertilizers, pesticides, and hybrid seeds has exerted significant pressure on the environment and has nearly compromised biodiversity (AWFDFFM, 2005: 2). To address these issues, AWFDFFM has implemented long-term programs that promote sustainable agriculture, provide assistance to groups of farmers and fishermen, and conduct capacity building, advocacy, information dissemination, and publication services.

Demonstrating equal partnership, collaboration, and local engagement is central to AWFDFFM's mission across Indonesia. The ideas of fishermen and farmers regarding their production priorities are deemed significant, and granting them autonomy is a crucial step toward democratizing the association. This democratization process is facilitated through active collaboration, dialogue, and progressive involvement with communities, the younger generation, and other civil society organizations, all aimed at enhancing alternative agricultural practices (AWFDFFM, 2005: 3).

A successful project requires the contracting party to perform in accordance with the agreed terms and conditions. According to the agreement between HIVOS Southeast Asia and AWFDFFM, the project formally commenced in June 2010, with AWFDFFM selected as the construction partner organization (CPO). One of the CPO's key functions, as outlined in the participation agreement ID074H38/PA-001, is to act as the authorized representative in the field during the period from June 21, 2010, to June 2011. Consequently, AWFDFFM must comply with the approved biogas design plans, which include predetermined sizes of 4 m^3, 6 m^3, 8 m^3, 10 m^3, and 12 m^3.

A business plan was designed to ensure that the construction partner organization (CPO) would effectively manage the number of biogas plants in alignment with their organizational capacity. In this context, AWFDFFM committed to constructing 50 biogas plants by June 2011 during its first year. AWFDFFM gained recognition for successfully installing around 70 biogas systems in Yogyakarta.

To support the establishment of these plants, an investment subsidy of IDR 2,000,000 was allocated to AWFDFFM for each biogas plant. Prospective owners were required to meet the following criteria: they must have at least 30 kg of cow dung (equivalent to three cattle) available at home; their location must not be prone to flooding or landslides; they must not have received any previous investment subsidies for a biogas plant; and the biogas plants must be constructed according to predetermined designs and standard operating procedures.

Additionally, an amount of IDR 100,000 was transferred to the CPO upon demonstrating successful maintenance of the biogas plants for a period of 12 months. In the event of issues such as a defective plant, these funds would be utilized to pay a third party for repairs. Meanwhile, the CPO would receive a fee of IDR 600,000 (approximately USD 60) for each biogas plant constructed. The areas of operation for the construction partner organization (CPO) are not geographically confined to a single regency; however, Bantul was selected as a key location due to AWFDFFM's strong familiarity and close affiliation with the local community. Additionally, the Yogyakarta-Central Java Assistance Program Round II in 2009 (YCAP II), which focused on improving organic farming, has facilitated access to the targeted community, as AWFDFFM has already established networks and recognition among the locals.

Before biogas construction can proceed, several procedures must be carried out for households (see Fig. 2.2). The initial step is to gauge the household's interest and commitment to adopting biogas. Following this, a survey is conducted to assess various factors, including the location of the house, the number of cattle or livestock owned, the number of household members, the readiness of the household to provide required construction materials, cooking activities, and sources of daily energy, as well as the availability of unoccupied land for the plant itself. Once the household has secured construction materials such as coral stones, bricks, sand, and cement, the construction process can begin. Prior to construction, the CPO will supply standard galvanized pipes, a manometer, and a biogas stove. On average, the construction process for one household takes 7–9 working days, depending on local climate conditions and the physical characteristics of the soil. Upon completion, a Biogas Plant Construction Contract (BPCC) will be signed by both the CPO and the household, serving as a legal document that confirms the CPO's commitment to upholding the quality of the biogas plant for a period of 36 months. According to the contract, the CPO is responsible for ensuring that the biogas plant maintains an average feeding rate of 100%. Therefore, intensive surveillance until 30 days after construction is a must to enable the users to feed in the plant until the gas is produced and blue flame on the gas-stove appears.

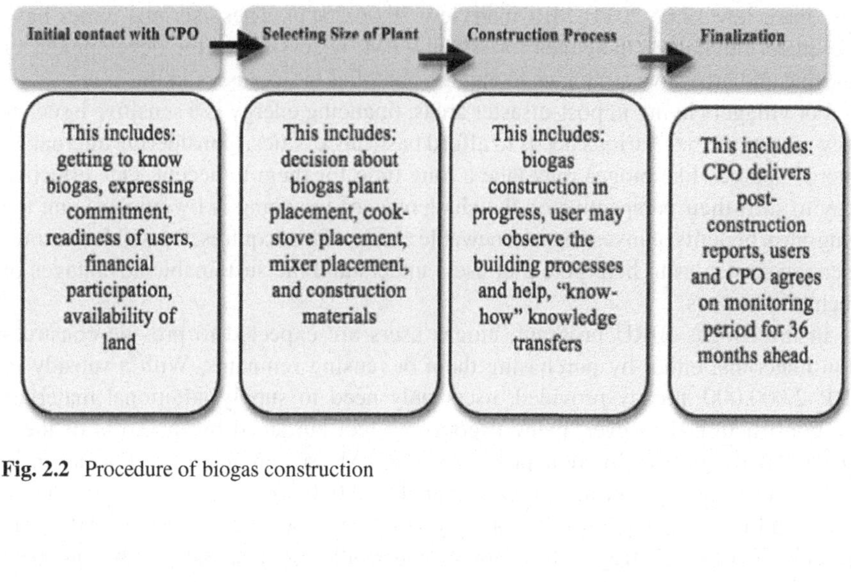

Fig. 2.2 Procedure of biogas construction

2.9 The Institutional Strategy in Reaching Out Prospective Users in the Region

In 2006, Bantul Regency, located in the southern part of the Yogyakarta Province of Indonesia, was severely affected by a 6.9 magnitude earthquake that devastated 75% of the area. Many people lost their families, and local businesses collapsed, leading to significant social stress and poverty. In this challenging context, the emergence of NGOs to provide aid and assistance was crucial for helping people sustain their livelihoods. Financial aid from the Indonesian government and international NGOs was granted to the community with the hope that residents would reconstruct their homes properly and create new businesses using available local resources. However, this aid inadvertently shifted the community's mindset from self-sufficiency to a dependence on external funding.

The Government of Indonesia launched a biogas project under the management of the Ministry of Energy and Mineral Resources, aimed at providing sustainable energy solutions for rural communities by offering biogas systems free of charge. This initiative was particularly important given the rising costs of kerosene and LPG, which had become unaffordable for many village residents. However, in Bantul Regency, the project faced significant challenges. Insufficient needs assessments to gauge user willingness, a lack of training programs, and inadequate quality assurance management resulted in many biogas installations becoming obsolete. Consequently, users abandoned their gas stoves and reverted to using kerosene stoves (Wl: interview 01.07.2011).

Moreover, villagers expressed reluctance to consider biogas offers from any parties, as they felt "traumatized" by the government's previous efforts, which had led to the occupation of their land for large installations (WI: interview 01.07.2011; TN: interview 17.06.2011; MR: interview 29.06.2011). These serious issues have presented inevitable challenges for AWFDFFM, necessitating the development of feasible and attractive strategies to engage potential biogas users in the area.

For villagers living in post-disaster areas, financing energy is a sensitive issue, as they must prioritize various needs to afford basic necessities. Considering alternative energy sources like biogas may take a long time for them to accept. One effective way to shift their perspective on the challenges of financing is by emphasizing the long-term benefits of investing in renewable energy. This requires accessible communication strategies to help potential users understand the sustainable advantages of such investments.

In the Biogas BIRU program, biogas users are expected to provide construction materials, either by purchasing them or reusing remnants. With a subsidy of IDR 2,000,000 already provided, users only need to supply additional materials for construction. However, many users often feel burdened by the costs of these materials. To address this reluctance, AWFDFFM uses an analogy: "For instance, if the construction materials are valued at IDR 2,000,000 to build a 4 cubic meter plant, and the biogas plant lasts for 30 years, that means your IDR 2,000,000 can provide 30 years of biogas. In contrast, you would incur no savings if you spent that money on LPG. Let's say a 20 kg LPG cylinder costs IDR 75,000 each month; therefore, IDR 2,000,000 would only cover 26 months of LPG consumption" (TN: interview 17.06.2011). This analogy helps bridge the gap in understanding the value of budgeting for construction materials, making the monetary implications clearer. A Biogas Partner Organization (BPO) adds,

> Rural farmers have limited access to education. Although biogas is closely related to methane gas and its impact on our atmosphere, such climate discussions are rarely understood by them. Using analogies enables them to grasp that IDR 2,000,000 is a small amount of money that can last for 30 years. In reality, biogas is much cheaper than they might think. (MR: interview 29.06.2011)

As AWFDFFM commits to establishing targeted households for the household biogas project ('Biogas Rumah'), a strategic approach is essential to achieve the goal of 75 biogas plants in Yogyakarta Special Province. Although biogas technology is relatively new to the organization, AWFDFFM has developed several strategies to engage potential users. One key strategy is assigning a knowledgeable staff member, skilled in both biogas technology and community engagement. This expertise is crucial, as the absence of qualified personnel could jeopardize the initiative's success (YK: interview 22.06.2011).

The long-standing trust AWFDFFM has built with the community, stemming from its involvement in various programs since the 1990s, has established the organization's credibility. Currently, in Bantul Regency, Yogyakarta Special Province, there are 50 biogas users, a number made possible through AWFDFFM's existing network in organic farming within the region. Rather than holding scheduled socialization sessions—whether monthly or weekly—to introduce the benefits of biogas, the organization relies on informal outreach methods to increase user participation. Two AWFDFFM officers mentioned that they did not allocate a specific budget for this project, as it was not included in their annual planning (YK: interview 22.06.2011; TN: interview 17.06.2011). From their perspective, it seems reasonable that there have been no set schedules for community outreach and socialization.

In Bantul, the activities of AWFDFFM are closely linked to local farmers organized under an organic farming program. During meetings, AWFDFFM takes the opportunity to educate farmers about the benefits of biogas for environmental sustainability and cow shed hygiene (TN: interview 17.06.2011). This approach has fostered established trust among the networks involved, which has proven to be a successful strategy for promoting biogas projects. With the assistance of the Biogas Promotion Officer (BPO), AWFDFFM can also reach communities that have not yet been exposed to their initiatives. The current BPO believes that a direct community approach is more effective for reaching potential users than public promotions through local newspapers. This strategy was innovated in November 2010, as she noted, "Previously, I knew that promotions were published in local newspapers and shared with local government, but the number of users wasn't growing. I believe this is because village residents have less time to read newspapers; they spend most of their time in the fields. So, I work with AWFDFFM to socialize programs at the grassroots level" (MR: interview 29.06.2011).

Attending local events, such as community gatherings (*perkumpulan*) of farmers in Sleman and Bantul, has become an invaluable opportunity for AWFDFFM staff to communicate the technical aspects and benefits of biogas. During my fieldwork, a group of students from Gadjah Mada University engaged AWFDFFM staff to present on biogas, aligning the topic with organic composting practices that were relevant at that time. Before the socialization began, the AWFDFFM staff member (TN) initiated the presentation with a friendly conversation with all attendees. He introduced himself in Javanese, creating an immediate connection. Without the aid of PowerPoint, he delivered an oral presentation outlining the benefits of biogas, including family savings, the use of residue as organic fertilizer, subsidies from HIVOS, and the advantages of having cleaner cow sheds. He communicated the details of biogas construction in simple, accessible language, avoiding technical jargon, and provided his contact number for anyone interested in learning more. Surprisingly, by the end of the meeting, more than ten attendees expressed their desire to have their homes surveyed for biogas installations. This instance demonstrates that an informal approach can be an effective promotional strategy. In my view, the secret to this successful socialization lies in the staff's ability to recognize the social context of their audience and use local language, which fosters a deeper cultural connection.

However, financing biogas remains a sensitive issue. Many people may react negatively or feel disappointed when confronted with the costs associated with installation. As elaborated above, users are likely to be reluctant to adopt biogas technology due to the perceived high cost of construction materials, which many deem unaffordable. This perspective could hinder efforts to increase the number of biogas users in Bantul Regency. Given that most participants in the promotional activities are farmers, it is understandable that they might hesitate to invest in construction materials, especially considering their low and often volatile monthly incomes, which lead them to think twice about such expenditures.

AWFDFFM views this situation as a challenge that can be addressed through effective communication strategies. A further assessment conducted by AWFDFFM in Bantul revealed that remnants of building materials (such as bricks, wood, pipes, and steel bars) from the rehabilitation of houses damaged in the 2006 earthquake are still readily available. According to TN's interview on June 17, 2011, many of these materials remain in the vicinity of local residences. Additionally, AWFDFFM's observations indicated that in Sri Hardono, where they established a demonstration plot, 90% of these remnants were still stored in the backyards of residents (YK: interview 22.06.2011). Leveraging this information, AWFDFFM emphasizes during promotional activities that reusing these remnants can significantly reduce costs compared to purchasing new materials. Furthermore, this approach not only alleviates financial burden but also contributes to a cleaner and healthier environment around the households.

The government's failure to effectively install biogas systems for rural communities poses significant challenges that require a different approach. In the context of biogas adoption, prospective users are often reluctant to register as customers. In rural Bantul, most individuals learn about inadequate biogas installations through fellow farmers, community meetings, and daily interactions within the village (MR:

interview 29.06.2011; TN: interview 17.06.2011). The government's lack of capacity in conducting needs assessments and providing technical knowledge further exacerbates the situation. For instance, Biogas Biru was once requested to repair a 6 m^3 biogas plant in one of Bantul's villages but declined due to financial constraints faced by the local government (WL: interview 01.07.2011). This situation has prompted AWFDFFM to develop an evidence-based approach aimed at countering the stereotype that biogas projects are inherently unsuccessful and redundant.

Building trust with prospective users is a top priority for AWFFDFFM, especially as they frequently question the credibility of the "Biogas Biru" program compared to government biogas initiatives. One effective strategy recently employed by AWFFDFFM is to invite current biogas users to socialization meetings, where they can share their daily experiences, the costs incurred for construction materials, and their overall satisfaction with the services provided by AWFFDFFM (YK: interview 22.06.2011). If attendees express interest in pursuing biogas installation during these meetings, AWFFDFFM arranges visits to ongoing biogas construction projects in Bantul. This allows prospective users to observe the projects firsthand and engage in dialogue with current biogas owners. As a result of these efforts, the number of biogas users increased significantly over a seven-month period, from June 2010 to January 2011 (YK: interview 22.06.2011).

Making novel technology attractive requires a presentation style that aligns with the audience's social context, especially when most prospective users come from farming backgrounds. In efforts to introduce biogas to the community, government representatives often focus on the greenhouse gas (GHG) effects that contribute to atmospheric damage. However, this approach has proven ineffective for local farmers, as it may create "cognitive blocks" due to their limited exposure to climate change issues. Consequently, this disconnect poses a barrier for the government in engaging the community with the program. As noted by MR in an interview on June 29, 2011, this approach does not resonate with the thinking of local farmers, who often have little knowledge of climate change. AWFFDFFM suggests that a more effective way to communicate the benefits of biogas is by presenting tangible, relatable information. For instance, explaining that 20 kg of cattle dung can be converted into biogas, sufficient for 25 years of cooking and replacing kerosene, makes the concept much more accessible to farmers (YK: interview 22.06.2011). Since its establishment in 1992, AWFFDFFM has formed a group focused on uniting organic farmers in Bantul, Indonesia, to accelerate the dissemination of organic farming values. This group promotes organic rice production and its distribution throughout the province. Although this group was robust in the organic movement until the 2000s, some members continue to hold monthly gatherings.

In the biogas project, AWFFDFFM has engaged this group with the hope that they can further disseminate the ecological mission of biogas to other farmers. From an ecological perspective, the bio-slurry or residue produced by biogas is rich in nutrients, which positively impacts plant growth. Organic farmers, in particular, have a strong ability to communicate the benefits of biogas from this standpoint (YK: interview 22.06.2011). Promoting biogas is flexible, and highlighting the advantages of organic compost derived from biogas residue is often more appealing to farmers than discussing technical details (MR: interview 29.06.2011).

References

Ahmed, Z., & Wang, Z. (2019). Investigating the impact of human capital on the ecological footprint in India: An empirical analysis. *Environmental Science and Pollution Research, 26*, 26782–26796.

Akinbami, J.-F. K., Ilori, M. O., Oyebisi, T. O., Akinwumi, I. O., & Adeoti, O. (2001). Biogas energy use in Nigeria: Current status, future prospects and policy implications. *Renewable and Sustainable Energy Reviews, 5*(1), 97–112. https://doi.org/10.1016/s1364-0321(00)00005-8

Ali, S., Yan, Q., Razzaq, A., Khan, I., & Irfan, M. (2022). Modeling factors of biogas technology adoption: A roadmap towards environmental sustainability and green revolution. *Environmental Science and Pollution Research, 30*, 11838–11860. https://doi.org/10.1007/s11356-022-22894-0

Alisjahbana, A. S., & Busch, J. M. (2017). Forestry, forest fires, and climate change in Indonesia. *Bulletin of Indonesian Economic Studies, 53*(2), 111–136.

Anderson, R., Bates, L., Johnson, E., & Morris, J. F. (2015). Packed bed thermal energy storage: A simplified experimentally validated model. *Journal of Energy Storage, 4*, 14–23.

Apergis, N., Polemis, M., & Soursou, S. E. (2022). Energy poverty and education: Fresh evidence from a panel of developing countries. *Energy Economics, 106*, 105430.

Bicker, A., Ellen, R., & Parkes, P. (2003). *Indigenous enviromental knowledge and its transformations*. Routledge.

Bijker, W. E., & Law, J. (1992). *Shaping technology/building society*. MIT Press.

Bößner, S., Devisscher, T., Suljada, T., Ismail, C. J., Sari, A., & Mondamina, N. W. (2019). Barriers and opportunities to bioenergy transitions: An integrated, multi-level perspective analysis of biogas uptake in Bali. *Biomass and Bioenergy, 122*, 457–465. https://doi.org/10.1016/j.biombioe.2019.01.002

Bresnihan, P., & Brodie, P. (2024). From toxic industries to green extractivism: Rural environmental struggles, multinational corporations and Ireland's postcolonial ecological regime. *Irish Studies Review, 32*(1), 93–122. https://doi.org/10.1080/09670882.2024.2304946

Budiman, I. (2021). The complexity of barriers to biogas digester dissemination in Indonesia: Challenges for agriculture waste management. *Journal of Material Cycles and Waste Management, 23*(5), 1918–1929.

Budiman, I., Muthahhari, R., Kaynak, C., Reichwein, F., & Zhang, W. (2018). Multiple challenges and opportunities for biogas dissemination in Indonesia. *Indonesian Journal of Energy, 1*(2). https://doi.org/10.33116/ije.v1i2.25

Budiman, I., & Smits, M. (2020). How do configuration shifts in fragmented energy governance affect policy output? A case study of changing biogas regimes in Indonesia. *Sustainability, 12*(4), 1358. https://doi.org/10.3390/su12041358

Buditama, A. (2016). Blue carbon for reducing the impacts of climate change: An Indonesian case study. *Environmental and Planning Law Journal, 33*(1), 68–88.

Callon, M. (1999). Actor-network theory—The market test. *The Sociological Review, 47*(1_suppl), 181–195. https://doi.org/10.1111/j.1467-954x.1999.tb03488.x

Callon, M., & Latour, B. (1992). Don't throw the baby out with the bath school! A reply to Collins and Yearley. In *Science as practice and culture*. University of Chicago Press.

Campbell, C. (1995). Conspicuous confusion? A critique of Veblen's theory of conspicuous consumption. *Sociological Theory, 13*(1), 37–47. https://doi.org/10.2307/202004

Capra, F. (2003). *The hidden connection*. Flamingo.

Carrier, J., & Miller, D. (1999). *From private virtue to public vice* (pp. 24–47). Polity Press.

Chan, C. C., Chan, F. C., & Tu, D. (2015). Energy and information correlation: Towards sustainable energy. *Journal of International Council on Electrical Engineering, 5*(1), 29–33.

Clemmensen, T. (2021). Human work interaction design for socio-technical theory and action. In *Human work interaction design* (pp. 11–50). Springer.

Crewe, E., & Axelby, R. (2013). *Anthropology and development: Culture, morality and politics in a globalised world*. Cambridge University Press.

Devadas, V. (2001). Planning for rural energy system: Part II. *Renewable and Sustainable Energy Reviews, 5*(3), 227–270. https://doi.org/10.1016/s1364-0321(00)00015-0

Do, T. H., & van Noordwijk, M. (2023). Accelerating subnational deforestation and forest degradation reduction efforts (REDD+): Need for recognition of instrumental and relational value interactions. *Current Opinion in Environmental Sustainability, 64*. https://doi.org/10.1016/j.cosust.2023.101330

Dugasseh, F. A., & Andersen, M. S. (2024). Non-carbon benefits of REDD+ implementation and sustainable emission reductions—A review. *Forests Trees and Livelihoods, 33*(4), 1–20. https://doi.org/10.1080/14728028.2024.2383737

Enrici, A., & Hubacek, K. (2016). Business as usual in Indonesia: Governance factors effecting the acceleration of the deforestation rate after the introduction of REDD+. *Energy, Ecology and Environment, 1*, 183–196.

Hartono, D., & Maharani, J. (2021). The impact of biogas utilization on poverty in Indonesia. *Jurnal Perencanaan Pembangunan: The Indonesian Journal of Development Planning, 5*(2), 230–249.

Henriques, J. J., & Schnorr, W. G. (2010, April 1). *Sustainability assessment and implementation of a biogas digester system in Western Kenya*. IEEE. https://doi.org/10.1109/green.2010.5453784

Herrmann, T., Jahnke, I., & Nolte, A. (2021). A problem-based approach to the advancement of heuristics for socio-technical evaluation. *Behaviour & Information Technology*, 1–23.

Izadi, N., Saadi, H., & Kooshki, L. (2024). Analysis of smallholder farmers' dynamics of knowledge sharing, skill transfer, and participation in using biogas (application of social network analysis). *Sustainable Futures, 8*, 100271. https://doi.org/10.1016/j.sftr.2024.100271

Kumar, S. (2016). Assessment of renewables for energy security and carbon mitigation in Southeast Asia: The case of Indonesia and Thailand. *Applied Energy, 163*, 63–70.

Kurniawan, R., & Managi, S. (2018). Economic growth and sustainable development in Indonesia: an assessment. *Bulletin of Indonesian Economic Studies, 54*(3), 339–361.

Lemma, B., Ararso, K., & Evangelista, P. H. (2021). Attitude towards biogas technology, use and prospects for greenhouse gas emission reduction in southern Ethiopia. *Journal of Cleaner Production, 283*, 124608.

Livingstone, S. (2003). On the challenges of cross-national comparative media research. *European Journal of Communication, 18*(4), 477–500. https://doi.org/10.1177/0267323103184003

Moore, W. E. (1963). But some are more equal than others. *American Sociological Review, 28*(1), 13. https://doi.org/10.2307/2090452

Mulyono, Y., Suranto, S., Yamtinah, S., Sarwanto, S., & Jamil, A. (2024). Biogas: Strengthening green energy infrastructure for a more sustainable future. *Journal of Biotechnology and Natural Science, 4*(1), 15–37. https://doi.org/10.12928/jbns.v4i1.10266

Murdiyarso, D., Lilleskov, E., & Kolka, R. (2019). Tropical peatlands under siege: The need for evidence-based policies and strategies. *Mitigation and Adaptation Strategies for Global Change, 24*, 493–505.

Nachmany, M., Fankhauser, S., Davidová, J., Kingsmill, N., Landesman, T., Roppongi, H., Schleifer, P., Setzer, J., Sharman, A., Singleton, C. S., Sundaresan, J., Townshend, T. (2015). *The GLOBE climate legislation study: A review of climate change legislation in 99 countries*. The Grantham Research Institute on Climate Change and the Environment.

Ni, J.-Q. (2024). A review of household and industrial anaerobic digestion in Asia: Biogas development and safety incidents. *Renewable & Sustainable Energy Reviews, 197*, 114371–114371. https://doi.org/10.1016/j.rser.2024.114371

Ni, J.-Q., & Nyns, E.-J. (1996). New concept for the evaluation of rural biogas management in developing countries. *Energy Conversion and Management, 37*(10), 1525–1534. https://doi.org/10.1016/0196-8904(95)00354-1

Ozturk, I. (2015). Measuring the impact of energy consumption and air quality indicators on climate change: Evidence from the panel of UNFCC classified countries. *Environmental Science and Pollution Research, 22*(20), 15459–15468.

Pal, R. C. (2005). *People participation in the promotion of biogas technology: A case study from western Uttar Pradesh* (Vol. 2, pp. 1–20). A Quarterly Electronic Newsletter on Renewable Energy and Environment.

Pandyaswargo, A. H., Wibowo, A. D., & Onoda, H. (2022). Socio-techno-economic assessment to design an appropriate renewable energy system for remote agricultural communities in developing countries. *Sustainable Production and Consumption, 31*, 492–511.

Pansera, M., & Fressoli, M. (2021). Innovation without growth: Frameworks for understanding technological change in a post-growth era. *Organization, 28*(3), 380–404.

Parawira, W. (2009). Biogas technology in sub-Saharan Africa: Status, prospects and constraints. *Reviews in Environmental Science and Bio/technology, 8*(2), 187–200. https://doi.org/10.1007/s11157-009-9148-0

Purohit, P., & Kandpal, T. C. (2007). Techno-economics of biogas-based water pumping in India: An attempt to internalize CO_2 emissions mitigation and other economic benefits. *Renewable and Sustainable Energy Reviews, 11*(6), 1208–1226. https://doi.org/10.1016/j.rser.2005.09.003

Rachman, H. O., & Barus, L. S. (2019). Impact of car-free day on air pollution and its multifarious advantages in Sudirman-Thamrin Street, Jakarta. *International Journal of Geomate, 17*(62), 167–172.

Rahman, M. S., Majumder, M. K., & Sujan, M. H. K. (2021). Adoption determinants of biogas and its impact on poverty in Bangladesh. *Energy Reports, 7*, 5026–5033.

Relva, S. G., da Silva, V. O., Gimenes, A. L. V., Udaeta, M. E. M., Ashworth, P., & Peyerl, D. (2021). Enhancing developing countries' transition to a low-carbon electricity sector. *Energy, 220*, 119659.

Robinson, B. L., Clifford, M. J., & Selby, G. (2023). Towards fair, just and equitable energy ecosystems through smart monitoring of household-scale biogas plants in Kenya. *Energy Research & Social Science, 98*, 103007. https://doi.org/10.1016/j.erss.2023.103007

Ryan, F. A., & Vivekananda, F. (1993). *Finding new routes in old paths: Linking cultural needs to technical knowledge appropriate technology inspires developing societies concept, controversy, and clarification.* Bethany Books.

Sealey-Huggins, L. (2017). "1.5°C to stay alive": Climate change, imperialism and justice for the Caribbean. *Third World Quarterly, 38*(11), 2444–2463. https://doi.org/10.1080/01436597.2017.1368013

Sforna, G. (2019). Climate change and developing countries: From background actors to protagonists of climate negotiations. *International Environmental Agreements: Politics, Law and Economics, 19*(3), 273–295.

Sharma, H. (2022). Mapping the global EdTech revolution during the pandemic: From 'determinism' to 'solutionism'. *Re-imagining Educational Futures in Developing Countries,* 119–137.

Shove, E. (2017). What is wrong with energy efficiency? *Building Research & Information, 46*(7), 779–789.

Tacconi, L. (2016). Preventing fires and haze in Southeast Asia. *Nature Climate Change, 6*(7), 640–643.

Talevi, M., Pattanayak, S. K., Das, I., Lewis, J. J., & Singha, A. K. (2022). Speaking from experience: Preferences for cooking with biogas in rural India. *Energy Economics, 107*, 105796.

Thompson, P., & Laaser, K. (2021). Beyond technological determinism: Revitalising labour process analyses of technology, capital and labour. *Work in the Global Economy,* 1(1–2), 139–159.

Thynell, M. (2018). Urban inequality in a fragile global city: The case of Jakarta. In J. Hellman, M. Thynell, & R. van Voorst (Eds.), *Jakarta* (pp. 15–40). Routledge.

Vijitharan, S., Tsusaka, T. W., & Sasaki, N. (2024). Evaluating REDD+ progress and results-based incentives in Sri Lanka: A comparative analysis across Southeast and South Asian countries. *Trees Forests and People, 16*, 100574–100574. https://doi.org/10.1016/j.tfp.2024.100574

Walekhwa, P. N., Mugisha, J., & Drake, L. (2009). Biogas energy from family-sized digesters in Uganda: Critical factors and policy implications. *Energy Policy, 37*(7), 2754–2762. https://doi.org/10.1016/j.enpol.2009.03.018

Wibowo, A., & Giessen, L. (2015). Absolute and relative power gains among state agencies in forest-related land use politics: The Ministry of Forestry and its competitors in the REDD+ Programme and the One Map Policy in Indonesia. *Land Use Policy, 49*, 131–141.

Wilhite, H. (2005). Why energy needs anthropology. *Anthropology Today, 21*(3), 1–2. https://doi.org/10.1111/j.0268-540x.2005.00350.x

Williams, N. B., Tilley, E., & Kalina, M. (2024). From failure to fairness: A call for accountability within household biogas development. *Energy Research & Social Science, 115*. https://doi.org/10.1016/j.erss.2024.103633

Winther, T., & Wilhite, H. (2014). An analysis of the household energy rebound effect from a practice perspective: Spatial and temporal dimensions. *Energy Efficiency, 8*(3), 595–607. https://doi.org/10.1007/s12053-014-9311-5

Chapter 3
Domestication of Biogas Technology

Abstract This chapter highlights the domestication theory to illustrate how biogas technology is integrated into the everyday lives of household users. This theory suggests that before implementing biogas technology in residential settings, individuals must possess specific knowledge, skills, and technical expertise, along with various tactics to find its appropriate "domain" within the household. Consequently, this chapter aims to help readers understand the socio-cultural nuances essential for making this technology acceptable to families. The discussion begins with the social construction of technology (SCOT) perspective, which posits that technology can drive significant social transformations and plays a dominant role in facilitating societal advancement. Through this lens, this study emphasizes the importance of considering the viewpoints of actors or users, who significantly influence how technology is designed based on their social values. The chapter explores the four steps of domestication: appropriation, objectification, incorporation, and conversion, detailing how each stage contributes to the successful integration of biogas technology in households.

Keywords Domestication · Biogas · Appropriation · Objectification · Incorporation · Conversion · Household users · Social construction of technology

3.1 Biogas in Social Construction of Technology Framework

Biogas as a home technology requires the harmonization of consumers with the technology itself. Before biogas technology can be effectively utilized in households, users must possess specific knowledge, skills, and technical know-how. Additionally, they need to adopt various strategies to facilitate its implementation in their homes. In reality, family life and energy consumption practices are deeply intertwined with the socio-cultural context, where each community member contributes to shaping ideas and introducing technology into the household. Similarly, the home serves as a site where societal values, conventions, and the envisioned future of society are

M. Alam, *The Social Life of Biogas*, https://doi.org/10.1007/978-981-96-3300-5_3

constructed (Wilhite, 2008). In line with this argument, the family should be viewed as a space where ontological stability is established, functioning as an economic entity that generates and reproduces products and services (Berker et al., 2005). The participation of households in defining economic activity impacts not only society but also individual lives, leading to a more nuanced conversation than is often assumed. To navigate the complexities of technology consumption, a suitable framework that incorporates socio-cultural components is essential, even though such frameworks are rarely linear. In this context, the concept of technology domestication serves as a valuable tool for understanding how technology is sociologically integrated into daily life. Therefore, this chapter will guide the reader in comprehending the framework and its specific elaborations, emphasizing the importance of socio-cultural nuances in making technology acceptable within families. In addition, we invite readers to engage in an enlightening conversation about the adoption of new technologies by emphasizing the socio-cultural complexities surrounding users.

The thesis of technology domestication stems from the social construction of technology (SCOT) approach, which emerged in the 1960s as a reaction against technological determinism. While technological determinism posits that technology drives social change and progress, SCOT highlights that users and social actors play crucial roles in shaping technology according to their needs and norms. This perspective underscores that technology does not exist in isolation; instead, it is co-shaped within the social context, providing a more nuanced understanding of how technology is integrated into society. The establishment of the social construction of technology (SCOT) theory emerged as a reaction against technological determinism. Engineering scholars who adhere to technological determinism posit that "hardware and software technology are the ultimate cause of social change" (Hughes, 1993: 8). This perspective overlooks the social relationships surrounding technology, as experts and engineers often exclude the societal factors embedded in users' world-views. Furthermore, technological determinism fails to account for "other factors inherent in the sociocultural setting" (Hughes, 1993: 16) in which technology operates. In contrast, SCOT introduces a fresh perspective by emphasizing the importance of social context in understanding technological adoption and change.

The proponents of the social construction of technology argue that technology is a socio-technical product shaped by "the conditions of its creation and use" (Wajcman, 2004: 34). The situated artifacts within the users' environmental milieu are co-shaped by their interactivity and the prevailing social context. Furthermore, the interpretation of these artifacts relies heavily on the users themselves. Scholars highlight that the knowledge and experience of users are two critical factors that can influence the co-shaping of technology (Bijker et al., 1989, cited in Omrod, 2006: 43; Mackenzie & Wajcman, 1999). In the social shaping of technology, the concept of interpretative flexibility is central, capturing how users' innovations arise from their diverse interpretations. Complementarily, the concept of the technological frame helps explain the development of heterogeneous socio-technical processes, thereby rejecting the notion of social reductionism. Moreover, Everett M. Rogers's Diffusion of Innovations theory falls short of fully addressing the complexities of technology adoption, particularly regarding socio-cultural dimensions.

A brief discussion of the social construction of technology (SCOT) is complemented by an exploration of the domestication of technology theory, including its principles and structure. Domestication signifies a shift from the earlier belief in the unidirectional transformational potential of technology, a perspective that dominated the diffusionism viewpoint until the mid-1980s, thus addressing a significant knowledge gap. Although the framework—comprising appropriation, integration, and conversion—may appear linear, it is, in reality, more complex. Domestication involves dialectical and reciprocal stages. This framework focuses specifically on identifying the socio-cultural aspects that govern the technical integration within the family context. Consequently, the emphasis lies on understanding the social and cultural characteristics inherent in each step of appropriation, integration, and conversion.

3.2 Social Construction of Technology as Roots of Domestication of Technology

The social construction of technology (SCOT) emerges as a counter-theory to the enduring concept of technological determinism, which positions technologists as the primary architects of societal development while sidelining the roles of society in shaping its community. This deterministic view neglects the myriad social dimensions, including cultural factors, that influence technological development. A key critique of technological determinism is its assertion that changes in technology are the most significant drivers of societal transformation. From this perspective, technology is perceived as an independent actor capable of instigating social change, thereby overlooking the complex interplay between technology and societal influences. However, the era of technological determinism did not endure, as the negative impacts of technology on society became increasingly apparent. Initially, many were captivated by the impressive successes of technologists and their innovations. Yet, public sentiment shifted toward discontent as users began to disengage from the technology. In light of this, the notion that technologists are the sole actors responsible for making technology available has become untenable in practical applications. This perspective underscores the importance of engaging all stakeholders in the establishment of technology, placing users at the forefront of the planning process to foster more harmonious interactions between users and technology. This heterogeneous relationship exists within a relational context, as the survival of modern society is rooted in the intricate harmonization of socio-technological systems that collaboratively support daily livelihoods.

The emergence of technological discourse has captivated sociology as a focal point for analysis, emphasizing how technology is interpreted within its social context. Sociology seeks to understand how society constructs technology according to its socio-cultural framework, prompting scholars to develop an intersectional approach rooted in constructionism as an overarching paradigm. Consequently, a new sociology of technology has emerged, focusing on in-depth studies of invention and innovation from an action-oriented perspective embedded in symbolic interactionism and post-structuralism (Law, 1992). This perspective, often referred to as "constructionism," analyzes technical change through concepts such as networks, translation, interpretative flexibility, and closure (Bijker et al., 2009). Constructionism links power as a primary driver of knowledge creation aimed at optimizing institutional authority, which in turn fosters innovation. This power is legitimized through rationality to control technology, positioning users—advocated by constructionism—as crucial participants in society who practice and re-develop new technological models. This new model allows users to play a significant role in solidifying negotiations essential for the sustainability and continuity of technology use.

The social construction of technology (SCOT) perspective and approach is fundamentally grounded in the understanding that society plays a crucial role in shaping existing science and technology. This perspective is often reiterated as the "Social Shaping of Technology," emphasizing that society acts as a determining force in the formation of technology. The establishment of this perspective can be attributed to two French philosophers and technological sociologists, Bruno Latour and Michel Callon, who first presented and utilized it as an analytical tool in the 1980s. However, as Fuglsang (2001: 39) notes, its origins date back to the early 1970s. By the 1980s, the term "Social Construction of Technology" had gained widespread recognition and emerged as a significant perspective for examining the interactions between users and technology, marking a dramatic scientific breakthrough during that period.

As previously discussed, technological determinism represents the conventional view of the relationship between science, technology, and society, asserting that technology exerts control over society. In contrast, the social construction of technology sheds light on the limitations of technological determinism by proposing that society and technology are equal and mutually influential, co-shaping one another. While both perspectives offer distinct values, Table 3.1 outlines key differences that help distinguish these concepts (Bijker, 2009: 22–24).

The social construction of technology (SCOT) approach, which emerged in the 1980s, serves as an embryonic concept that paved the way for the later development of domestication theory. Central to SCOT are the concepts of "interpretative flexibility" and "technological frame," each possessing distinct epistemological value in understanding how society comprehends technological artifacts and the contexts in which these technologies are situated according to the social groups that engage with them.

"Interpretative flexibility" posits that different social groups associated with a technology can ascribe varying meanings to the same technological artifact. In this regard, the social variables in discussions about technology become sociological factors that highlight how the use of technology is profoundly influenced by the intrinsic identities of users, including their knowledge, norms, beliefs, and perceptions. Consequently, the relationship between users and technology can be described as a "seamless web" (Fuglsang, 2001). The second key concept is the "technological

Table 3.1 Distinction between standard view of science, technology & society and constructionist view of science and technology and society

Standard view of science, technology, and society	Constructionist view of science and technology and society
There is clear distinction between the empirical and the scientific/technical domain	Both domains are intertwined; what is defined as a technical or as a political problem will depend on the particular context
Difference between "real science" and transciences	All science is value-laded and nay again depending on the context-have implications for regulation and policy; thus there is no fundamental difference between "real science" and "trans-science," "mandated sciences," or "policy relevant science"
Scientific knowledge is discovered in the making of technology and to make it sound direction, which are answered ambiguously by nature	The stabilization of scientific knowledge is social process
There are real responsibilities of scientists as well as technologist, which is a key issue	Development of science and technology is asocial process rather than a chain of individual decisions political and ethical issues related to science therefore cannot be reduced to the question of social responsibility of scientist and technologist
Technology develops linearly for example inception → decision → operation	Technology development cannot be conceptualized as a process with separate stages, let alone a linear one
Distinction between technology's development and its effects	The real construction of technology is a process that also continuous into what is commonly called its "diffusion stage; the (social, economic, ecological, cultural …) effects of technology are thus part of the construction process and typically have direct vice versa implications for technology's shaping"
Clear distinction between technology development and control	Technology does not have the context-dependent status that is necessary to hope for a separation of its development and control its social construction and the (political, democratic) control are part of the same process
Clear distinction between technology stimulation and regulation	Stimulation and regulation may be distinguishable goals but need not necessarily be implemented separately
Technology determines society, not the other way around	Social shaping of technology and technical building of society are two sides of the same coin
Social needs as well as social and environmental costs can be established unambiguously	Needs and costs of various kinds are also socially constructed-depending on the context, they are different for different relevant social groups, varying with perspective

frame," which is proposed to elucidate the heterogeneous socio-technical ensembles and to define the intersections between the "actors" within a relevant social group. This concept emphasizes that a new technological artifact is not only shaped by a specific culture but is also constructed through a combination of enabling and constraining interactions among the relevant social groups in distinct ways. Furthermore, the "technological frame" is linked to broader concepts of institutions and social structures (Fuglsang, 2001). According to Bi, this framework scrutinizes the concept of power, suggesting that society, armed with its knowledge, has the capacity to change or modify technological artifacts. The social construction of technology (SCOT) approach profoundly influences the social interactions between users and technological artifacts (Fuglsang, 2001). Two core conceptual perspectives—interpretative flexibility and the technological frame—are essential for exploring users' interdependence in shaping the meaning of biogas technology in their daily lives.

Users play a crucial role in determining and shaping these artifacts. Their ability to discern the meaning of the artifacts lies in interpretative flexibility, which asserts that changes in technology are not fixed; they evolve and are redefined during implementation. Wajcman (2004: 37) defines interpretative flexibility as the "capacity embodied in the group of users or people engaged in a technology," noting that understanding varies among users, allowing them to create alternative meanings and uses for technology. Different social groups may attribute alternative meanings to technology, shaped by their unique social experiences, which is where the characteristic of "heterogeneity" comes into play. Heterogeneity refers to the differences in the meaning-creation processes embedded in technology. As Volti (2001: 75) articulates, it represents a state of "meaning-making" generated by the actors involved in utilizing technology, allowing them to choose and adapt the technology according to their "needs, desires, and intentions." Experiences thus form the foundational basis for distinguishing the conceptualization of artifacts (Flichy, 2007: 8). Here, innovation emerges as a set of actions that may follow the meaning-making process, potentially reshaping the artifacts. Akrich (2001) elaborates that this dynamic can lead to new knowledge creation, asserting that technology possesses the unique ability to generate and naturalize new forms and orders of causality and indeed, new forms of knowledge about the world.

The *technological frame* addresses the interactions formed between actors within social groups after the adoption or use of a technology. This concept helps analyze how post-adoption behaviors emerge, showing how "existing practices guide future practices, though not completely determined" (Bijker, 2009: 27). It highlights that the continued use of a technology is often shaped by subsequent learning processes among users and collaborative efforts. While interactions may face constraints, these challenges allow for the optimization of the technology. Bijker (2009) succinctly states that "technology is constructed by a combination of enabling and constraining interaction within relevant groups in a specific way" (p. 27).

User-driven innovation, which stems from interpretative flexibility, is significant because users are integral actors in the socio-technical regime. Several studies on SCOT affirm that users are the primary source of innovation, often improving and extending the capabilities of technological artifacts (Flichy, 2007; Ornetzeder &

Rohracher, 2006; Rohracher, 2001, 2003). As key contributors to the sustainable use of artifacts, users influence innovation through their expectations, social practices, and even institutional roles (Ornetzeder & Rohracher, 2006: 140). Exploring these factors in the context of innovation reveals dynamic practices, usage patterns, and co-shaping processes between technology and users.

Artifacts and users are intertwined, with users playing a crucial role in rationalizing technology through their experiences. However, the role of designers cannot be separated from this process. Flichy (2007) argues that negotiations between users and designers are essential, as users often interpret artifacts differently. While engineering perspectives may view this as contradictory, it creates opportunities for mutual learning. Rohracher (2003) expands on this by noting that the mutual learning between designers and users can occur on a broader scale within technological regimes. Moreover, the roles, responsibilities, and perceptions of users can further sustain innovation (Geels, 2010).

In the context of biogas technology, where the artifact is situated within an agricultural society, local knowledge and user experiences are essential elements for household users. As Escobar (1999: 9) notes, local agricultural knowledge should be viewed in the context of innovation as "context-specific improvisation" rather than merely as indigenous knowledge. This perspective implies that innovation arises from users adapting to the specific settings or spatial contexts where the technology is employed. Through experimentation with the artifacts, local modeling can emerge, leading to new practices (Escobar, 1999: 9). Cultural values are deeply intertwined with local practices, influencing how innovation unfolds. These values can shape how objects are ordered and how users experience and behave in relation to the technology. Innovation is an integrative process, reformulating design in ways that are co-shaped by values, experiences, ideas, norms, and local practices. This dynamic space often lacks clear boundaries between users and technical experts, both of whom contribute to the development of the artifacts, enabling continuous innovation (Rohracher, 2003). Independence in operating the technology is a crucial outcome, highlighting users' capabilities to maintain its ongoing functionality. For instance, in biogas development, when technical issues arise, users are expected to troubleshoot and repair the reactor independently, finding alternative solutions to persistent problems. This knowledge transfer within society fosters independence and self-reliance.

In the biogas construction process, the role of the actor is represented by the construction partner organization, such as AWFFDFFM, a non-governmental organization (NGO). NGOs contribute by building trust and mobilizing public support for advancing innovation. As Rohracher (2009) emphasizes, the NGO's role in "negotiating standards, enrolling and aligning supply-side and demand-side actors, communicating with the public, and building trust for the respective products" is crucial for ensuring that user innovations are less conflicting and more acceptable to various groups.

The role of users in innovation and in making sense of new technologies within the household and community is elucidated through the concept of the domestication of technology. As new technological artifacts emerge and are adopted within households, each household exhibits distinctive patterns in integrating these innovations into their established daily practices. Households are not isolated entities; rather, they are inclusive spheres where ideas about artifacts can evolve through the process of domestication. As Silverstone et al. (2006) argue, the household functions as a "dynamic space," actively engaging with the public world of production and the exchange of commodities and meanings. The domestication process, while an inevitable trajectory within families, can vary significantly depending on the social context.

Domestication, as articulated by Blechar et al. (2006), refers to the process through which new technological artifacts are integrated into daily routines, transforming them into familiar elements of everyday life. This concept underscores the dynamic interactivity between users (actors) and technology, illustrating how the adoption of an alien artifact reshapes both routines and the artifact itself. In this sense, domestication becomes central to the social shaping of technology, as it highlights how users actively negotiate and redefine technology's role within their lives. The incorporation of technology into domestic settings involves not just the assimilation of the artifact but also the evolution of social practices, values, and identities surrounding it, thereby illustrating the reciprocal relationship between society and technology. In each household, the ways of incorporating and adapting to new technology can vary significantly. Nevertheless, certain typical structures or processes are involved in domestication, as elaborated by Berg (2006: 99), which include appropriation, objectification, incorporation, and conversion.

Firstly, appropriation refers to the moment when users obtain or attain the technology, often characterized by a distinctive act of ownership as the artifacts arrive at home. Secondly, objectification pertains to the changes in behavior or actions that reflect users' tastes, values, and styles. Thirdly, incorporation describes how technological artifacts are integrated into the routines of everyday life. Finally, conversion delineates the process in which household members strive to immerse the technology within their existing values and societal context. The concept of domestication is closely linked to the social shaping of technology; as technology evolves, so too do the everyday practices surrounding it (Aune, 1996, cited in Christensen, 2009: 434).

Furthermore, the appropriation process of technology inevitably requires users to grasp the meaning of what the artifact truly represents. This concept is what Selwyn (2003: 108) refers to as "sense." She elaborates that when a user encounters a specific technology or artifact, they imbue it with their own "lived experience or life world" to make it more meaningful and beneficial in their lives. The construction of meaning around each artifact is considered one of the key factors that significantly shape these experiences. The functions of artifacts may instigate certain objectives and symbolically encode meanings, which significantly contribute to the phases of domestication (Hynes & Rommes, 2005; Mackay & Gillespie, 1992; Pantzar, 1997; Sørensen, 2006). Identity also plays a crucial role, as it signifies the different social statuses that users may later identify with (Sørensen, 2006: 47). The production of meaning and social identity is inextricably linked to the broader cultural context in which the artifact exists. Thus, the concept of domestication emerges from the above approach, inspired by the social shaping of technology and serving as a lens for analysis. As highlighted by Hynes and Rommes (2005), through the concept of domestication, users play a dominant role in "defining the nature, scope, and function of the technology itself" (p. 149). The mundane routines within households illustrate that the production of meaning for artifacts requires a process of interpretation, which allows the concept of interpretive flexibility to find its place in the domestication of technology. According to Ward, domestic routines are the key to this concept, as he explains, "There is no technology without action, and [this] implies a degree of interpretative flexibility when users attempt to integrate the technology into their domestic routines" (p. 149). From this perspective, domestication is born from the social shaping of technology and aligns with the social construction of technology framework.

In many perspectives, the adoption of technology is widely regarded as the primary approach to understanding how certain technologies come into the hands of users. However, this perspective often fails to account for the significance of culture in the adoption process. In response to this limitation, the domestication approach emerges as a means to move beyond the belief in the one-sided transformative power of technology. This approach broadens the study of innovation and diffusion to include cultural and social worldviews of specific social groups.

Evert M. Rogers' perspective on technology adoption typically concludes once technological artifacts arrive in users' homes. Yet, the ways in which technology is practiced within the household space remain underexplored. When technology enters the home and becomes integrated into daily routines, it necessitates a different approach from Rogers' diffusionism. The concept of technology-in-use emphasizes an action perspective, highlighting micro-relations in everyday settings where we engage with technologies—or, more specifically, technical artifacts—by incorporating and using them (Sørensen, 2006).

Furthermore, we are shaped by these artifacts, which influence our routines and prompt us to engage with them, becoming "occupied" by their modified forms, usable functions, and capabilities. The harmonization, integration, synchronization, and even the seamless relationship between humans and technology are the outcomes of the domestication process (Silverstone et al., 1989, in Lie & Sørensen, 2006: 10). Regarding the socio-cultural integration of biogas technology, users draw upon their own understandings, influenced by their subjectivities. This form of rationality encompasses various sources for thought, referred to as a "field of rationality." The main premise of field rationality stems from the understanding that societal structures can significantly influence energy consumption in households. These structures encompass ideas, values, norms, conventions, and codes that facilitate social interaction at various levels within social groups. Individuals often rely on these elements to inform their preferences, actions, and emotional responses. Consequently, structure embodies culturally specific ways in which people think, behave, and act.

These societal elements constitute the tools through which individuals engage with their environment—shaping how they think, act, and feel. Additionally, the availability of natural resources, material objects, regulations, and formalized procedures also inform choices, decisions, and everyday energy usage. Furthermore, the capacity of humans as resources should not be overlooked. This context allows us to learns that bodily strength and knowledge are important features that make the use of renewable equipment workable. In any given society, conventions and values guide practices related to material management. These conventions are formed by the resources possessed by the group, which include both human resources (knowledge, capacity, competencies) and material resources (e.g., the availability of shops in the neighborhood, household income, and the type of dwelling). Biogas technology is utilized by cohabiting families, where sociological dynamics play a significant role. Family members often share certain norms and values that may align with or contradict those encountered outside the household. These external norms can influence individual behaviors within the home. The values, norms, and preferences held by each family member shape the negotiation process regarding the use of technology. While households and groups are significantly influenced by external structures and broader social networks, they also possess the capacity to negotiate and establish their own practices based on internal power relations, common aspirations, and available resources.

Figure 3.1 illustrates four main factors across three levels—society, group, and individual demonstrating the relationships among human resources, non-human resources, and norms and values.

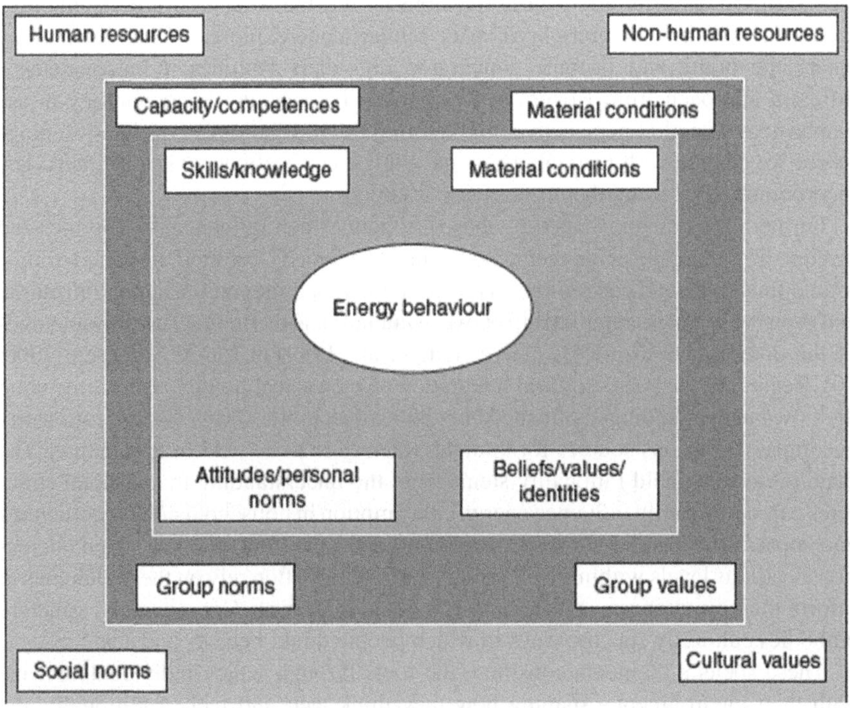

Fig. 3.1 This image is a diagram illustrating the factors influencing energy behavior, categorized under human and non-human resources

3.3 Domestication of Technology

Domestication refers to the process of taming something unfamiliar, a term commonly used in animal husbandry when wild animals are adapted to serve the needs of a local community. This concept, however, has broader applications. Thomas Berker, a sociologist from the Norwegian University of Science and Technology (NTNU) in Trondheim, extends the term to technology. He explains that domestication involves integrating a new and often unfamiliar technology into the diverse environment of a household. These "strange" or "wild" technologies must be adjusted and harmonized with the household's existing conditions. The primary goal is to ensure that the technology aligns with the social structures, daily routines, and values of the household members, enabling them to adopt and adapt to it seamlessly. Before technology is fully integrated into a household, individuals who are unfamiliar with it or have had little prior experience with it often resist its adoption. If this resistance persists, the technology may eventually become obsolete due to a lack of use. To prevent this, the domestication of technology approach is essential. Successful domestication is evident when users' perceptions shift—when the technology is no longer seen as "cold, lifeless, or problematic" but rather as a functional and reliable tool that enhances daily life.

Different scholars may offer varying interpretations of domestication, but they generally agree on its significance. One common thread is the pivotal role of users in this process. According to Sørensen (2006), the successful domestication of technology ultimately depends on its continued use by the users themselves. He emphasizes that users actively shape patterns of use, which stem from the deeper meanings they associate with the technology (Sørensen, 2006: 45). The domestication of technology outlines three key aspects: routines in its use, the role artifacts play in shaping identities, and the cognitive processes involved. As Sørensen (2006: 47) explains, domestication involves constructing a set of practices related to an artifact. This includes routines in using the artifact and the development of institutions to support and regulate its use. The second aspect relates to the meaning of the artifact, particularly its role in the production of users' identities. The third aspect addresses the co-learning process, which highlights how the cognitive process of adapting technology is deeply intertwined with culture and habituation, changing users' understanding over time. Sørensen (2006) also emphasizes that as users modify their practices, the meaning of technology evolves—not just its functionality. From this perspective, domestication is a process in which artifacts become tied to practices, meaning, users, and other artifacts, forming complex socio-technical networks (Sørensen, 2006: 47). This complexity is echoed by De Laet and Mol (2000), who describe it as the "fluidity of technology," reflecting the dynamic nature of these networks. The views of Sørensen and Berker align on the importance of changing practices. At this point, it is learnt that users co-create their environment and technology, reshaping artifacts to meet their evolving needs. Domestication, therefore, is about adapting technology to fit the shifting demands of users, households, and workplaces.

Another expert, Lehtonen (2003), views domestication of technology as a lens to closely examine the unique practices each household adopts. When new technology is introduced into the family and combined with existing technologies, it creates new practices, often referred to as a "technoscape." Lehtonen (2003: 364) explains that domestication provides insight into how people shape their technoscapes by integrating new technologies into their lives. This technoscape is shaped by the dynamic exchange between production and everyday use within the family, where meaning-making plays a critical role in how households engage with technology. Lehtonen also notes that the relationship between families and technology often becomes affectionate. This emotional connection forms because users actively maintain a process of co-learning, not only with the technology but also within the user network itself. Domestication, then, is not simply about one-sided control—either humans over technology or vice versa—but about creating an emotional and reciprocal relationship between people and things (Lehtonen, 2003: 381).

Disruptions in the use of technology can result from these emotional dynamics, and such disruptions often lead to new knowledge. The integration of new technologies creates a shared space that helps make technology more acceptable within the family. However, conflicts can arise when the introduction of technology disregards the physical boundaries of the household. Lehtonen suggests that these "trials and struggles" (2003: 381) are an inevitable part of normalizing the presence of technology in the home.

Symbolic meaning plays a crucial role in the domestication process, as it arises from the coexisting ideas, local values, and societal norms surrounding the users. While studies on symbolic meaning within domestication vary, many scholars agree that it is shaped by cultural entities, social systems, and the collective ideas of a society (Hynes & Rommes, 2005; Selwyn, 2003). Symbolic meaning can be understood as the societal perception of technology, which imbues the user with distinct identities. For example, biogas, as a technological artifact, may serve to highlight the user's social status (Bourdieu, 1997), given its high cost, which is unaffordable for many within the community. In this context, Hynes and Rommes (2005) argue that the symbolic meaning of an artifact is rooted in the social capital of the community. They contend that participation in social networks, along with access to resources unavailable to other community members, helps shape the symbolic meaning of the artifact. The ability of users to purchase certain technologies by leveraging resources such as money, land, or family support in the form of contributions is seen as an achievement. Such actions can lead to shifting values, where users move from a focus on material wealth to an emphasis on environmental consciousness—reflecting changes in their value orientations. This process of achievement, disposition, and value transformation is a key aspect of symbolic meaning production, as it ties technology to the cultural capital of the society (Hynes & Rommes, 2005). Selwyn (2003) similarly notes that symbolic capital emerges at every stage of technological use, influencing the way the technology is perceived and incorporated. According to Berker et al. this symbolic meaning exists within the broader cultural context. One of the most intriguing insights from Hynes and Rommes (2005) about spatial separation within the "homescape" of users. Using television as an example, they found that users often place this technology in rooms reserved for household members. Guests are typically excluded from these spaces unless they share an intimate relationship with the family. This case illustrates how symbolic meaning often involves a separation between private and public spaces in the domestication of technology.

The evolution of technology extends beyond its physical or technical aspects; it also encompasses its usage, which is closely tied to the knowledge possessed by users. When users are well-informed and have access to relevant information, innovation and modification can emerge as natural extensions of this knowledge development. Materiality and technical possibilities are key factors that users often consider when determining whether such innovations can be negotiated. Lynn et al. (1996) have studied the ways in which these elements can influence the domestication of technology. According to their research, the knowledge embedded within users' internal sources significantly shapes the extent of possible innovation. This process is influenced by the interdependencies between materials, which ultimately dictate the technical possibilities of the objects.

3.4 The Framework of Domestication of Technology

The integration of technological artifacts into the household involves navigating the social and cultural complexities that exist within the family environment. This process often requires adjustments in knowledge, the mobilization of resources, and changes to daily practices to ensure that the technology is accepted by the household. Although a systematic discussion on how technological artifacts become embedded in households and the specific conditions required for their sustained use is seldom addressed, Berg (2006) argues for the importance of developing a framework to understand this process. He proposes a series of phases that outline how technology is introduced, adopted, and adapted by users in their homes. These phases are appropriation, objectification, incorporation, and conversion, each representing a distinct stage in the domestication of technology.

Domestication is the process of integrating various elements machines, ideas, values, and information into the home, which inherently involves crossing boundaries (Silverstone, 2005). Below are further explanations of the stages of technology domestication.

3.4.1 Appropriation

Before a technological artifact arrives home, it is generally available on the market, allowing people to navigate various choices for the specific technology they wish to acquire for their needs. This concept is closely related to how users and households can afford technological artifacts. Berg (2006) defines appropriation as the "moment of acquisition of the technology, including the act of ownership and the arrival of the new commodity in the home. This may involve buying the technology." In the context of households, the strategies employed to facilitate the purchase of technology are influenced by various motives, collective persuasion, and adequate exposure to relevant information. It is challenging to prioritize these factors to determine which ones come first or last in the decision-making process. Building on this concept, this research aims to explore how households develop specific strategies within their families and communities to access biogas technology.

Hynes and Rommes (2005) propose that appropriation involves the activities in which potential customers engage while searching for specific strategies to acquire a desired technological artifact. Customers' desires when selecting a technological artifact are aligned with their needs and may necessitate the involvement of external stakeholders who can provide adequate mediation for potential users. In this context, the broader social structure can influence potential users' inclination toward adopting household biogas technology.

Although the factors facilitating the adoption of biogas technology are broad, key elements include the social system, kinship, user needs, cultural context, dissemination of information, availability of locally obtainable materials, and community knowledge. These elements are essential for analyzing the context of this biogas technology case study. Warriner and Moul (1992) assert that kinship can significantly influence resource provision through exchanges, thereby accelerating adoption. Additionally, other scholars emphasize that culture and the constitution of knowledge are critical for enabling technology adoption among potential users in rural areas (Driesen & Popp, 2010; Grieve, 2004; Lawal, 2010). The need for using technology is another fundamental aspect that is integral to its acceptance within the household. In this context, culture and individual needs are closely intertwined. As one scholar argues, when individuals' needs for specific technologies are reflected in the tools they use regularly, the culture surrounding those technologies remains sustainable. These societal factors can significantly influence potential users, guiding them to make firm decisions and actions toward adopting biogas technology. To avoid any ambiguity in the concept of adoption, it is essential to utilize succinct definitions provided by scholars. For instance, Rogers (2003) defines adoption as a specific initial action in which individuals are sufficiently ready to embrace the technology (Subedi et al., 2009). However, Rogers (2003) challenges the notion of appropriation as merely an act performed by individuals when integrating particular technologies into their daily practices.

The possession of knowledge is crucial for enabling the adoption of appropriate technology. Subedi et al. (2009) argue that knowledge can significantly shape decision-making processes, ensuring that specific technologies align with users' needs and are perceived as useful innovations. In this context, knowledge encompasses users' understanding of various technical aspects related to biogas, including engineering processes, maintenance procedures, and routine operations necessary for its sustainability.

Studies conducted by various scholars indicate that the impact of knowledge interacts with other factors, such as financial capabilities, the ability to process information, and the meaning-making process of artifacts within the context of adopting biogas technology (Cudjoe Bensah & Brew-Hammond, 2010; Subedi et al., 2009; Tripp, 2001). Another important aspect to consider is the necessity of equal partnerships among all involved stakeholders. Given that local people and their contexts are at the forefront of this initiative, forging partnerships must fundamentally respect and appreciate the traditions and customs of the communities involved. In essence, transparency among all parties is vital to the partnership. According to Tripp (2001), in pursuing effective and ideal renewable energy solutions, public partnerships should embrace transparency, equity, and environmentally responsible technology transfer to local populations.

Establishing strong partnerships and securing government support can significantly help overcome financial barriers to adoption. Addressing funding challenges for the construction process is often a critical obstacle. In this regard, government assistance is essential for users to navigate these financial constraints. Moreover, government commitment to supporting and promoting biogas technology, in collaboration with NGOs, can greatly enhance efforts to boost adoption rates (Katuwal & Bohara, 2009). For example, the Nepal Alternative Energy Promotion Center (AEPC) successfully aimed to establish 100 biogas plants in Nepal by 1997, with endorsement from the Ministry of Science and Technology to accelerate the adoption of biogas technology.

In the adoption process, the availability of information regarding the technical aspects of the technology significantly influences decision-making. This information must be tailored to the context in which potential users reside. The more comprehensive and relevant the information, the greater the interest it will generate among potential users. Using local language and context in communication makes the information more accessible, thus encouraging wider adoption. It is crucial to note that providing digestible information through examples from local settings is particularly effective in fostering successful adoption. From this perspective, we learn that participatory communication is crucial for helping potential users comprehend the technology in a language they understand. In this context, information about biogas technology is disseminated through socialization within the local community to enhance awareness regarding financial considerations, basic technical maintenance, construction aspects, and potential sustainability challenges. Pal (2003) conducted an in-depth study on the impact of socialization in introducing biogas technology to community members. His primary finding indicates that users can alleviate their doubts and confusions about the perceived risks associated with biogas by participating in socialization stages, which are often complemented by trial activities that allow them to experience the technology firsthand.

In addition to socialization and information dissemination, it is crucial for potential users to test the technology to determine whether it meets their needs and expectations. The rationale behind this is that hands-on experience with the technology before adoption enables potential users to better understand its functionality (Rogers, 2003). Moreover, trials of biogas operations provide practical experience in addressing technical issues related to daily use, such as operating the stove and checking the manometers. In Lusaka, Zambia, the demonstration of biogas plants has proven successful in engaging more adopters. These events not only generated interest but also helped mitigate users' dissatisfaction with the new technology.

At the appropriation stage, potential users may have distinct preferences regarding which model of biogas technology to adopt, making this preference a foundational aspect of the adoption process. These preferences are grounded in the suitability of the technology to meet the users' needs, aligning with the felt needs of potential users. When individuals recognize their felt needs, they are more likely to deeply consider the purpose of the technology in their lives, rather than viewing it as a mere short-term solution. If the chosen renewable energy aligns with the users' needs, they will perceive an urgency to implement the technology.

Motivation plays a crucial role as a personal state that can empower potential users to reinforce their commitment to adopting biogas technology. It encourages decisive action toward the technology. Furthermore, motivation derived from personal willingness can drive users to operate the chosen technology and deepen their understanding of its technical aspects. In the context of biogas adoption, motivation, when supported by adequate local capacities, effectively enables users to integrate biogas into their daily routines (Pal, 2003).

Additionally, intrinsic motivation encourages users to engage in behaviors that ensure the continuous functioning of the biogas technical system. Information about post-construction services can further facilitate knowledge acquisition, enabling users to develop independent strategies for addressing potential challenges. Kumari and Grover (2007) argue that stronger household motivation, particularly in response to rising oil prices, can lead to decisive actions in transitioning to biogas technology.

3.4.2 Objectification

Once a technological artifact is purchased or agreed upon for placement in a chosen area of the home, households often have specific preferences regarding its location. These preferences for placement and installation are inseparable from the pre-existing values of the household. This aspect relates to the concept of "objectification" within domestication theory. According to Berg (2006), objectification is defined as "the way the household or its individual members express their own values, tastes, or style through their display of the new technology." Departing from this concept, this research explores how biogas technology is installed in households and the extent to which pre-existing values, tastes, habits, and social norms influence these preferences. Blechar et al. (2006) coined the importance of the inclusion of resources and other cultural capitals. They coin that the mobilization of material resources, skills, networks, and family connections can significantly contribute to the success of the objectification process. These elements serve as modalities that users can leverage during the initial phase of installing a technological artifact.

The stage of objectification is crucial as it determines how effectively users adapt the technological artifact for their needs. As highlighted by Berg (2006), the functionality of technological artifacts is enhanced when users can attach meaning to them, ensuring they operate efficiently. Knowledge acquisition during the appropriation process is vital for enabling users to utilize the technological artifact effectively. The types of knowledge required may vary based on the complexity of the technology and the skill levels involved in optimizing its use. Novak and Gowin (1984) identify two critical types of knowledge for developing countries in assessing technology: "knowledge about technology" and "knowledge about attributes." The former, often referred to as technical knowledge or know-how, involves understanding how the technology works, while the latter pertains to individuals' perceptions and assessments of the product's quality.

Further discussion on know-how knowledge encompasses four sub-conceptions: (1) Knowledge of quality, (2) Knowledge of social and economic contexts, (3) Knowledge of ways to learn, and (4) Knowledge of ways to marshal and organize knowledge (Novak and Gowin, 1984). Firstly, knowledge of quality pertains to users' awareness of the quality of a certain technology. This aspect involves considering the availability of local materials and how they can be mobilized, the amount of capital (both human and financial) required for organization, the technical operations involved, and the procedures for addressing complaints regarding any technical glitches associated with the technological artifacts. Secondly, knowledge of social, cultural, and economic contexts emphasizes that technology exists within a societal framework that influences the daily interactions between users and technological artifacts. This type of knowledge encompasses modes of social and cultural learning, scrutinizing how knowledge is transferred when technology is introduced by designers or experts. Users must consider their pre-existing practices when starting to utilize newly adopted technological artifacts. In a closely related perspective, Westkog et al. identify three crucial aspects of societal context: beliefs, values, and identities. The cultural context is inherent in the use of technology; for instance, Bharadwaj et al. (2023) observed in her biogas study in India that users limited the use of biogas cookstoves for their daily routines because biogas could not achieve the necessary heating level to toast "dal." This cultural practice reflects local ingenuity and traditions. Thirdly, knowledge of how to learn involves the potential for innovation regarding the technological artifact, enabling users to adjust it to meet their evolving needs.

3.4.3 Incorporation

As technological artifacts are situated within the home, they are not isolated from pre-existing household practices. The ways in which household users negotiate and seek alternatives to integrate newly adopted technology are therefore of considerable importance. This process, referred to as **"incorporation"** in the context of domestication, involves examining "how technology is integrated into the routines of everyday life" (Berg, 2006). Blechar et al. (2006) asserts that incorporation encompasses the process by which technology becomes an integral part of our everyday routines. Household biogas technology is utilized within the specific social systems of communities. The newly adopted artifacts may either complement or compete with existing energy sources, which are heavily influenced by community social activities. In this context, the social system serves as an embedding force that shapes the practice of technological artifacts (Altman & Wohlwill, 2012). Conversely, the social system can facilitate a meaning-making process regarding technology. The social system's involvement transforms appropriate technology into a tangible element that resonates with the worldview of local people, fostering their awareness (Capra, 2003). This heightened consciousness ultimately contributes to the sustainable functionality of the technology, as it becomes part of the users' rational approach to integrating technology into their lives. When artifacts undergo enhancement through specific processes of creation, kinship plays a crucial role in disseminating innovation (Bennett & Despres, 1960). Kinship can be envisioned as a supportive space where innovations are integrated, as Peletz (1995) suggests. Kinship acts as a conduit for information within the transferred group, serving as a "habitus" for symbolic assets and trust. Additionally, Altman and Wohlwill (1978) emphasize that the solidarity of kinship is deeply intertwined with political and economic relations, reflecting the motives of society and the dynamics of power among both users and non-users.

Other experts focus on the incorporation stage as one where technological artifact necessitate specific adjustments to existing user practices, a process often referred to as habituation. For instance, users may allocate time and designate specific spaces to ensure that the technological artifact is effectively integrated into their daily routines. In this context, Hynes and Rommes (2005) assert that for a technological artifact to become functional, it must secure a place within the routines of daily life. Everyday life is crucial for examining how technology influences changes within various spheres of the household. This aligns with the argument put forth by Lie and Sørensen, who assert that technology can potentially impact all aspects of life, including work, home, and leisure.

By investigating how technology co-shapes daily life, we can gain insights into the extent to which individuals can exert control over their lives. When a household successfully adjusts its pre-existing practices and routines to accommodate a new technological artifact, that artifact effectively finds its "place" in daily life, minimizing potential problems. The key to successful usage lies in the household's ability to make practical adjustments to their existing behaviors, allowing the artifact to seamlessly integrate into their living space. The use and acceptance of technological artifacts within a society are often linked to the social standing of households within local values. The interactions between community members and the perceptions of technology users can be influenced by the adoption of these artifacts, which may enhance their social status and bring added meaning to their lives. This aspect of the domestication process is referred to as "conversion." According to Berg (2006), the concept of conversion addresses the social changes that emerge within a household after technology becomes widely recognized in the community. It also involves how household members adapt the technology to align with their own values and societal perspectives.

Furthermore, during the process of conversion, changes in the household's practices involving both pre-existing and newly adopted technological artifacts become evident. Through this transition, household members begin to discover the benefits that the technology brings, often integrating it into their daily routines as a sustainable tool for supporting the household. The advantages that users derive from the artifact may vary depending on their basic needs and other factors influencing their usage. This leads to what scholars describe as "interpretative flexibility"—the idea that different users experience and interact with the technology in unique ways, shaped by their individual circumstances and needs, setting them apart from others who use the same artifact.

3.5 Conclusion

From the discussion of the theoretical perspective above, it is evident that the concept of technology domestication possesses distinctive characteristics that set it apart from other theoretical frameworks addressing technology adoption. This concept emphasizes the social and cultural dimensions involved in the acceptance of technological artifacts within the home. While the process of domestication may appear linear, it offers opportunities for a deeper exploration of how technological artifacts come into the users' hands and the subsequent social effects on their lives. The four distinct stages of domestication—appropriation, objectification, incorporation, and conversion—serve as a valuable framework for researchers to understand the life cycle of a technological artifact, from its initial introduction in the market through commodification to the transformation of values resulting from its integration into the community.

References

Altman, I., & Wohlwill, J. F. (1978). Introduction: The confluence of developmental and environmental psychology. In *Children and the environment* (pp. 1–4). Springer US.

Altman, I., & Wohlwill, J. F. (2012). *Human behavior and environment: Advances in theory and research* (Vol. 2). Springer Science & Business Media.

Akrich, M. (2001). De l'accusation à la cause. L'intégration des controverses et des conflits dans le processus d'innovation. In Goujon, P.et Heriard Dubreuil, B (Ed). *Technology and ethics, a European quest for responsible engineering* (pp. 157–167).

Bharadwaj, B., Kambo, A., Arratia-Solar, A., & Ashworth, P. (2023). Awareness of biogas increases its social acceptance. *Journal of Cleaner Production,* 421, 138432.

Bennett, J. W., & Despres, L. A. (1960). Kinship and instrumental activities: A theoretical inquiry. *American Anthropologist, 62*(2), 254–267. https://doi.org/10.1525/aa.1960.62.2.02a00050

Berg, A.-J. (2006). *Technological flexibility: Bringing gender into technology (or was it the other way round?).* Open University Press.

Berker, T., Hartmann, M., & Punie, Y. (2005). *Domestication of media and technology*. McGraw-Hill Education.

Bijker, W. E. (2009). *Social construction of technology: A companion to the philosophy of technology* (J. K. B. Olsen, S. A. Pedersen, & V. F. Hendricks, Eds.). Blackwell.

Bijker, W. E., Bal, R., & Hendriks, R. (2009). *The paradox of scientific authority: The role of scientific advice in democracies*. MIT Press.

Blechar, J., Knutsen, L., & Damsgaard, J. (2006). Reflexivity, the social actor, and m-service domestication: Linking the human, technological, and contextual. In *Designing ubiquitous information environments: Socio-technical issues and challenges. IFIP—The International Federation for Information Processing* (pp. 57–70). Springer EBooks. https://doi.org/10.1007/0-387-28918-6_7

Bourdieu, P. (1997). Selections from the logic of practice. In *The logic of the gift* (pp. 190–230). Routledge.

Capra, F. (2003). Deep ecology: Educational possibilities for the twenty-first century. *NAMTA Journal, 28*(1), 157–193.

Christensen, T. H. (2009). "Connected presence" in distributed family life. *New Media & Society, 11*(3), 433–451. https://doi.org/10.1177/1461444808101620

Cudjoe Bensah, E., & Brew-Hammond, A. (2010). Biogas technology dissemination in Ghana: History, current status, future prospects, and policy significance. *International Journal of Energy and Environment, 1*(2), 2076–2909. https://www.ijee.ieefoundation.org/vol1/issue2/IJEE_06_v1n2.pdf

de Laet, M., & Mol, A. (2000). The Zimbabwe bush pump. *Social Studies of Science, 30*(2), 225–263. https://doi.org/10.1177/030631200030002002

Driesen, D. M., & Popp, D. (2010). Meaningful technology transfer for climate disruption. *Journal of International Affairs, 64*(1), 1–15. https://www.jstor.org/stable/24385182

Escobar, A. (1999). After nature: Steps to an antiessentialist political ecology. *Current Anthropology, 40*(1), 1–30.

Flichy, P. (2007). *Understanding technological innovation: A socio-technical approach*. Edward Elgar.

Fuglsang, L. (2001). Three perspectives in STS in the public context. In *Visions of STS* (pp. 33–49). State University of New York Press.

Geels, F. (2010). *The role of cities in technological transitions: Analytical clarifications and historical examples*. Routledge.

Grieve, R. H. (2004). Appropriate technology in a globalizing world. *International Journal of Technology Management and Sustainable Development, 3*(3), 173–188. https://doi.org/10.1386/ijtm.3.3.173/0

Hughes, K. S. (Ed.). (1993). *European competitiveness*. Cambridge University Press.

Hynes, D., & Rommes, E. (2005). "Fitting the internet into our lives": IT courses for disadvantaged users. In T. Berker, M. Hartmann, Y. Punie, & K. Ward (Eds.), *Domestication of media and technologies* (pp. 123–144). Open University Press.

Katuwal, H., & Bohara, A. K. (2009). Biogas: A promising renewable technology and its impact on rural households in Nepal. *Renewable and Sustainable Energy Reviews, 13*(9), 2668–2674. https://doi.org/10.1016/j.rser.2009.05.002

Kumari, R., & Grover, I. (2007). Waste generated and adoption of waste management practices among rural households in Haryana. *Journal of Human Ecology, 22*(4), 355–360. https://doi.org/10.1080/09709274.2007.11906046

Law, J. (1992). Notes on the theory of the actor-network: Ordering, strategy, and heterogeneity. *Systems Practice, 5*(4), 379–393. https://doi.org/10.1007/BF01059830

Lawal, Y. (2010). Poverty alleviation and the use of appropriate technology. *International Journal of Academic Research, 2*(6).

Lehtonen, T.-K. (2003). The domestication of new technologies as a set of trials. *Journal of Consumer Culture, 3*(3), 363–385. https://doi.org/10.1177/14695405030033014

Lynn, L. H., Mohan Reddy, N., & Aram, J. D. (1996). Linking technology and institutions: The innovation community framework. *Research Policy, 25*(1), 91–106. https://doi.org/10.1016/0048-7333(94)00817-5

Mackay, H., & Gillespie, G. (1992). Extending the social shaping of technology approach: Ideology and appropriation. *Social Studies of Science, 22*(4), 685–716. https://doi.org/10.1177/030631292022004006

MacKenzie, D., & Wajcman, J. (Eds.). (1999). *The social shaping of technology* (2nd ed.). Open University Press.

Novak, J. D., & Gowin, D. B. (1984). *Learning how to learn*. Cambridge University Press.

Omrod, S. (2006). Lets nuke the dinner: Discursive practices of gender in the creation of a new cooking process. In C. Cockburn & R. F. Dilic (Eds.), *Bringing technology home: Gender and technology in a changing Europe*. Open University Press.

Pal, J. (2003). The developmental promise of information and communications technology in India. *Contemporary South Asia, 12*(1), 103–119. https://doi.org/10.1080/0958493032000123399

Pantzar, M. (1997). Domestication of everyday life technology: Dynamic views on the social histories of artifacts. *Design Issues, 13*(3), 52. https://doi.org/10.2307/1511941

Peletz, M. G. (1995). Kinship studies in late twentieth-century anthropology. *Annual Review of Anthropology, 24*(1), 343–372. https://doi.org/10.1146/annurev.an.24.100195.002015

Rogers, E. M. (2003). *Diffusion of innovations* (5th ed.). Free Press.

Rohracher, H. (2001). Managing the technological transition to sustainable construction of buildings: A socio-technical perspective. *Technology Analysis & Strategic Management, 13*(1), 137–150. https://doi.org/10.1080/09537320120040491

Rohracher, H. (2003). The role of users in the social shaping of environmental technologies. *Innovation: The European Journal of Social Science Research, 16*(2), 177–192. https://doi.org/10.1080/1351161030451 6

Rohracher, H. (2009). Transforming environmental innovation systems: Innovation cultures and social learning processes. In *Innovation cultures—Challenge and learning strategies* (pp. 69–91). Filosofia.

Rohracher, H., & Ornetzeder, M. (2006). *Socio-technical research in the context of, sustainable technology programmes* (pp. 1–20; IFZ-Electronic Working Papers).

Selwyn, N. (2003). Apart from technology: Understanding people's non-use of information and communication technologies in everyday life. *Technology in Society, 25*(1), 99–116. https://doi.org/10.1016/s0160-791x(02)00062-3

Silverstone, R. (2005) Domesticating domestication. Reflections on the life of concept. In T. Berker, M. Hartmann, M. Y. Punie and K. Ward (eds) *Domestication of media and technologies* (pp. 229–248). Open University Press.

Silverstone, R., Berker, T., Hartmann, M., Punie, Y., & Ward, K. (2006). *Domestication of media and technology*. Open University Press.

Sørensen, K. H. (2006). Domestication; the enactment of technology. In T. Berker, M. Hartmann, Y. Punie, & K. Ward (Eds.), *Domestication of media and technology*. Open University Press.

Subedi, M., Hocking, T. J., Fullen, M. A., McCrea, A. R., Milne, E., Wu, B., & Mitchell, D. J. (2009). An awareness-adoption matrix for strategic decision making in agricultural development projects: A case study in Yunnan province, China. *Agricultural Sciences in China, 8*(9), 1112–1119. https://doi.org/10.1016/s1671-2927(08)60319-3

Tripp, R. (2001). Agricultural technology policies for rural development. *Development Policy Review, 19*(4), 479–489. https://doi.org/10.1111/1467-7679.00146

Volti, R. (2001). *A STS perspective on technology and work*. SUNY Press

Wajcman, J. (2004). *Techno feminism*. Polity Press.

Warriner, G. K., & Moul, T. M. (1992). Kinship and personal communication network influences on the adoption of agriculture conservation technology. *Journal of Rural Studies, 8*(3), 279–291. https://doi.org/10.1016/0743-0167(92)90005-q

Wilhite, H. L. (2008). *Consumption and the transformation of everyday life: A view from South India*. Palgrave Macmillan.

Chapter 4
Setting of the Scene

Abstract This chapter addresses the research methods used to investigate household users' experiences and practices with biogas as an innovative technology. It also explores the social nuances, societal contexts, and local community experiences related to its use. The objective of this study is to examine the everyday use of biogas in home activities and social interactions among users, thereby enhancing its long-term sustainability and promoting its innovation. A qualitative research approach is employed as the primary method for extracting nuanced insights from the social environment. This approach includes interviews, observations, and participation in community events, enabling the author to delve into the subjective experiences, perspectives, and practices of household biogas users.

Keywords Biogas · Subjective experiences · Users · Societal setting · Everyday use · Innovation

4.1 Research Design

This chapter outlines the methods employed to study household users' experiences and practices with biogas as a novel technology. The research explores social nuances, societal settings, and local experiences, all imbued with interpretative meaning. A qualitative approach is deemed suitable for this study (Østern et al., 2023; Woods & Davids, 2023) as it allows the author to grasp the nuances of interaction through interpretative efforts. This includes observations, interviews, and various methods that capture the social contexts in which communities live (Creswell, 2005; DeWalt & DeWalt, 2002). Using a case study model within the qualitative approach, key factors influencing biogas adoption were examined in relation to the social context. This study focuses on the everyday use of biogas technology, specifically the successful adoption of household biogas in Sri Hardono village, Bantul, Yogyakarta. An intrinsic case study was selected, as it allows the author to deeply explore and better understand this particular case by examining it through the lens of the specific problem. Through this approach, the author investigated the successful integration of biogas

M. Alam, *The Social Life of Biogas*, https://doi.org/10.1007/978-981-96-3300-5_4

technology, analyzing how households socially construct and engage with it within their community.

Different methods of data collection are essential for case studies, as Creswell (2007: 123) notes, "a case study involves the widest array of data collection as the researcher attempts to build a depth picture of the case." In this study, the author seeks to explore the successful adoption of biogas technology in Sri Hardono village by examining it comprehensively within the societal context and settings. As highlighted in the introduction, the key factors the author aims to investigate include the dissemination of information among community members, kinship relations, and the use of locally available materials for biogas construction. These aspects are expected to play a significant role in the successful adoption of biogas within households. Additionally, the research scrutinizes how biogas technology is socially integrated into both the community and household levels. Indicators of successful adoption are explored through community factors such as the spread of information, local materials, kinship and interpersonal relations, and the incorporation of technological artifacts into daily life.

The decision to use a case study approach was further validated by a preliminary survey conducted in the village and an in-depth dialogue with AWFDFFM. During this stage, it became clear that Sri Hardono village exhibited unique characteristics that likely influenced the adoption of biogas technology. These include: (i) the open-mindedness and willingness of the village chief, who allowed his land to serve as a training center and demonstration plot for biogas construction in Yogyakarta province; (ii) AWFDFFM's assertion that the villagers maintain strong social bonds and family affiliations, which may facilitate the dissemination of information through community meetings, family conversations, and daily interactions, even for those who did not attend formal socialization sessions; and (iii) the villagers' exposure to previous organized activities, such as organic farming initiatives under YCAP-II by AWFDFFM, university research, and government interventions in promoting bio-slurry as an innovative fertilizer for paddy field production. These factors have likely acted as stimulators, fostering a growing knowledge base among users and contributing to the successful adoption of biogas as a novel technology. Secondly, the case study primarily aims to understand the process and local experiences, focusing on the specific and detailed aspects within a broader context (Ramachandra, 2008). The goal of this research is to explore the everyday experiences related to how biogas is integrated into household routines and how social interactions among users contribute to its sustainable operation and innovation as an alternative energy source. Additionally, the qualitative approach serves as a fundamental platform for capturing the subtle nuances of the social setting, allowing for a deeper understanding of how biogas is appropriated and adapted by the community.

4.2 Geography of the Study Area

Prior to the project launch in June 2010, AWFDDFM signed a contract with HIVOS Southeast Asia, partnering as the construction partner organization for the household biogas project, locally referred to as "Biogas Rumah." Sri Hardono was chosen as the pilot area not only for implementing the biogas initiative but also for serving as a demonstration plot and providing mason training. This strategic decision emphasized both the practical application of biogas technology and the skill development of local masons, ensuring the sustainability and scalability of the project.

In June 2010, Sri Hardono Village was selected for a pilot project on household biogas, organized by the AWFDFFM. The village was chosen due to its previous involvement in the Yogyakarta-Central Java Community Assistance Program (YCAP)-II, which focused on enhancing farmers' knowledge and skills in organic farming, improving agricultural business practices, and promoting health through the use of organic fertilizers. The program also aimed to build a network among farmer groups across the Yogyakarta Special Province, leveraging social capital as a key asset in the success of organic agriculture. In the Pundong district, two villages participated in the YCAP-II initiative: Sri Hardono and Panjang Rejo, both located near each other. In Sri Hardono village, AWFDDFM selected three sub-villages—Pranti, Jonggrangan, and Tangkil for participation in the household biogas pilot project. More than 50 farmers from both Pranti and Tangkil sub-villages became beneficiaries, while in Jonggrangan, 60 farmers participated. The project's primary long-term goal was to reduce the local community's dependence on chemical fertilizers, promoting healthier families and more sustainable farming practices. The selection of Sri Hardono village was also influenced by the strong networking among project participants, which facilitated information sharing and collaboration.

The study was conducted in a village located in the southern region of Bantul Regency, Yogyakarta Special Province, Indonesia, situated approximately 20 km from the capital city of Yogyakarta. The village is accessible via common transportation modes such as buses, cars, and motorcycles. The area is predominantly agricultural, with arable land, and comprises 17 sub-villages or *pedukuhan* in Javanese, which include Sawahan, Candi, Monggang, Tangkil, Pundong, Baran, Piring, Ganjuran/Kembangkerep, Seyegan, Nangsri, Klisat, Tulung, Gulon, Jonggrangan, Paten, Pranti, and Potrobayan.

The fieldwork for the biogas study was primarily conducted in Tangkil, Pundong, Gulon, Jonggrangan, and Kembangkerep sub-villages. Additionally, visits were made to informants in other villages, including Jetis, Karangasem, and Tirtomulyo, as these households had family affiliations with biogas users in the main study areas. These neighboring villages are about a 15-min drive from the capital of Bantul Regency. The selection of these areas was influenced by both geographical proximity and familial connections, which played a significant role in the dissemination of information about biogas technology and its adoption within the community. Geographically, Sri Hardono village is located in the intersecting point of two major rivers in Yogyakarta Special Province, which are Opak and Oyo Rivers and for years these rivers have been bringing advantages to the community as the locals can obtain sand and coral stone for housing constructions, but the local government have recently set up local regulation to limit the sand extraction. The morphological formation of the village is much affected by the dynamic fault lines of those two rivers since thousand years ago (BAPPEDA Bantul, 2009). The risk of flooding due to thick sedimentation of those rivers. However, until recent there have no disaster report unveiling about the higher possibility of flooding.

4.3 Socio-Economic Condition of Study Area

In the context of demography, it was reported that since 2009, 12,753 people have been inhabiting Sri Hardono village. Of this population, 49.20% are men, and 50.80% are women. This nearly equal gender distribution provides insight into the balance of the population in the village and forms the basis for understanding the social dynamics in various contexts, including labor force participation, community involvement, and household roles. Additionally, the population in Sri Hardono village can be categorized by productive manpower. Those aged 15–49 years old dominate the demographic, making up 56.91% (7,258 people), while 25.37% (3,236 people) of the population are aged above 50 years old. This indicates that the majority of the population is of working age, which may influence economic activities and labor dynamics within the community.

In terms of education, 33.44% of the population are graduates of senior high school, reflecting the level of formal education in the village. This educational background provides an understanding of the community's capacity for adopting and sustaining technological innovations, such as biogas technology. The number of households in each of the 17 sub-villages differs, and Table 4.1 presents the distribution of households across these sub-villages. This information is crucial for understanding the scale and scope of biogas adoption and its impact on community life.

For administrative purposes, the government of Indonesia classifies the prosperity rate of families in Sri Hardono village into different categories: privileged and less-privileged households. The National Ministerial Office for Population (BKKBN) defines privileged households as those capable of fulfilling all kinds of basic needs, including psychosocial needs, and who can continuously contribute to society while actively engaging in community activities. Families who are able to meet their basic needs but whose livelihoods remain isolated from other psychosocial needs, such as education, transportation, and social interaction, are classified as less-privileged households. In contrast, families who are unable to meet either their basic or psychosocial needs are categorized as poor households.

According to data provided by BAPPEDA Bantul (2009), the majority of households in Sri Hardono village fall into the less-privileged category, with 2,342 households (60.20%). The second largest group consists of poor families, totaling 875

Table 4.1 Households in each sub-village of Sri Hardono village

No	Sub village	Number of household	Total population	Number of household living in poverty
1	Sawahan	156	503	14
2	Candi	205	700	44
3	Monggang	298	1020	43
4	Tangkil	239	880	69
5	Pundong	335	1161	69
6	Baran	288	797	69
7	Piring	218	713	75
8	Ganjuran	226	733	58
9	Seyegan	248	783	43
10	Nangsi	202	614	65
11	Klisat	192	610	41
12	Tulung	217	618	48
13	Gulon	193	621	59
14	Jonggrangan	195	633	42
15	Paten	255	798	45
16	Pranti	228	891	52
17	Potrobayaban	206	674	57
	Total	3,841	12,754	875

Source BAPPEDA Bantul (2009)

households (22.78%). Privileged families make up the smallest minority in the village. The detailed breakdown is presented in Table 4.2.

Compared to other households in the village, those involved in this study can be classified as relatively privileged. They face no significant barriers in accessing basic needs for their family members and are capable of meeting additional requirements, such as transportation, education (with at least nine years of schooling, from primary to secondary high school), and participation in community meetings. A closer look at their homes reveals that the walls are cemented, and the floors are tiled with ceramic, further indicating that these households have good access to economic resources, which likely benefits them when purchasing construction materials.

The majority of local people work in the agricultural sector, comprising 28.93% (1,752 people), with 19.98% (906 people) specifically working as farmers (BAPPEDA Bantul, 2009). This data highlights the continued dominance of agriculture in the village. Cows, in particular, have traditionally been viewed as cultural symbols in the community. As Nuradita and Subiyanto (2024) explain, the Javanese term "Rojo Koyo" derives from *Raja* (king) and *Kaya* (wealth), signifying that cows represent high social status. However, cows are not typically raised for productive purposes like milk or meat production to support the family's daily economy. Instead,

Table 4.2 Prosperity rate of households in each sub-village of Sri Hardono village

No	Sub village	Level of prosperity			Total
		Privileged households	Less-privileged households	Poor households	
1	Sawahan	24	118	14	156
2	Candi	38	123	44	205
3	Monggang	23	232	43	298
4	Tangkil	59	11	69	239
5	Pundong	38	228	69	335
6	Baran	13	140	75	228
7	Piring	20	140	58	218
8	Ganjuran	44	139	43	226
9	Kembangkerep	58	125	65	248
10	Nangsi	37	124	41	202
11	Klisat	26	128	48	192
12	Tulung	37	134	59	217
13	Gulon	88	63	42	193
14	Jonggrangan	19	131	45	195
15	Paten	61	142	52	255
16	Pranti	29	142	57	228
17	Potrobayaban	20	135	51	206
	Total	624 (1.25%)	2,342 (60.20%)	875 (22.78%)	3,841 (100%)

Source BAPPEDA Bantul (2009)

they are often sold in times of financial need, such as during costly life events like a son's circumcision (*khitanan*), paying educational fees, or other major expenses.

In the aftermath of the 2006 earthquake, the rural economy collapsed as local markets were devastated, leading to a 10% increase in the village's poverty rate. To revive the local economy, the government, in collaboration with local NGOs, launched soft loans (*pinjaman lunak*) for those interested in establishing small and medium enterprises (SMEs). These initiatives have since contributed to the local economy by creating jobs and empowering the village workforce.

4.4 Local Traditions

All locals in the village are Javanese and speak Javanese as their mother tongue. The term "Javanese" refers not only to the cultural identity of the people but also to their households, signified by the prominence of the Javanese language in daily communication. It also represents their broader cultural heritage, including their descent from those who have historically inhabited Java Island (Susilo & Syato, 2016). Communal work, such as road construction and village infrastructure maintenance (*sambatan*), remains a tradition. On weekends, groups of youths and adults practice Javanese traditional music (*cokekan*), performing at wedding celebrations. Weekly Qur'an recitals (*pengajian*) are held for both men and women in each sub-village. According to Ayuwandani and Tondok (2023), these community-led events serve as platforms to maintain social cohesion among Javanese villagers. Despite the predominantly Muslim population, there has been no significant political or social upheaval among Catholics, Christians, and Muslims, as social cohesion is maintained through shared community activities and events. The social values of Javanese in upholding harmony is in accordance with the study undertaken by Lestari et al. (2013: 33) as follows:

> the concept of harmony in the perspective of Javanese has similarities with the concept of social harmony in East Asian society. The difference is that the Javanese perspective puts more emphasis on togetherness, hoping for a united harmony, while the East Asian society puts emphasis on the achievement of balance of harmonious value and instrumental harmony. Harmony that is achieved by maintaining connectedness will encourage individuals to minimize conflicts in order to avoid negative judgments from social environment. Harmony that is achieved by maintaining compatibility and harmony itself will also encourage individuals to minimize conflicts to avoid disintegration.

On Figs. 4.1 and 4.2 Yogyakarta Special Province is located in the Java Island (marked with red circle) Source: http://www.lib.utexas.edu/maps/middle_east_and_asia/indonesia.gif

Fig. 4.1 Map showing Southeast Asia and parts of Oceania, highlighting Indonesia. A red circle marks the city of Yogyakarta on the island of Java

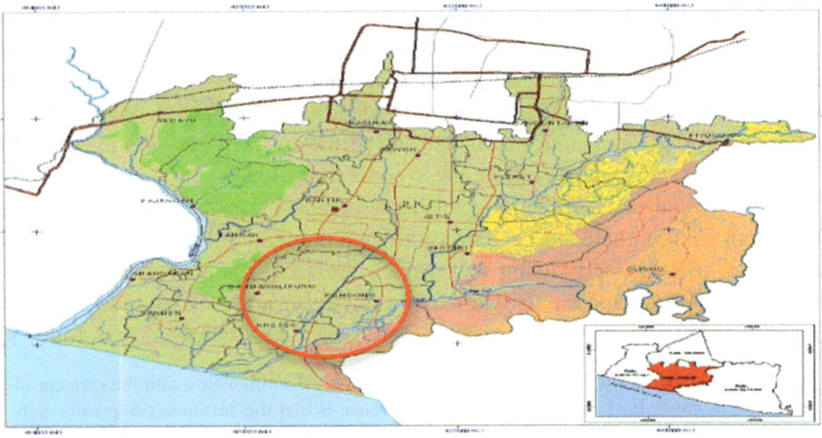

Fig. 4.2 Location of Bantul Regency within Yogyakarta Province. Study area is marked with "red circle" (*Source* jogjainvest.jogjaprov.go.id, http://www.lib.utexas.edu/maps/middle_east_and_asia/indonesia.gif)

4.5 Collecting of Data

4.5.1 Personal Construct

This research aims to explore the social phase of biogas technology appropriation in rural households. By investigating how technological artifacts integrate into the daily lives of users, it also seeks to uncover the meanings users ascribe to this technology, which ultimately leads to its acceptance by family members (Creswell, 2005). A qualitative approach has been chosen for this study, as it is well-suited to understanding individual experiences within a nuanced context. Previous studies have highlighted the importance of focusing on household experiences, as they reflect human relationships and lived realities within the home (Berker et al., 2005; Hall, 2003). Experience is characterized by an individual's attitude in responding to specific stimuli or situations within a given social context. Based on this rationale, the "Personal Construct" approach was applied in this study to capture individual experiences related to acquiring and integrating biogas technology into the household space. The personal construct theory considers how individuals interpret and structure their phenomenological world (Hall, 2003). From this perspective, the study explores and examines users' perceptions of the technology, seeking to understand what the technology offers them and how it influences their lives.

Additionally, the personal construct theory originates from phenomenological sociology, which views the everyday world as a realm where individuals can engage and transform their realities through their actions and interactions (Bakardjieva, 2006: 70). Furthermore, comparing different individuals' constructs regarding specific artifacts is essential for understanding the shared experiences that shape their everyday lives. In practice, engaging with phenomenology is quite intriguing. Before delving deeper into its application, Berker et al. (2005: 7) suggest that allowing participants or family members to use the technology while reflecting on the shared responsibilities between husband and wife can effectively support the social construct approach. Considering this, an ethnographic examination is conducted, focusing on the facts, thoughts, perceptions, and contextual materials surrounding the technology (Sullivan & Brockington, 2003). In this context, the meaning derived from technology is emblematic, and as McCarthy and Wright (2004) advise, we must endeavor to "combine what culture and language have crystallized from the past with our current feelings, desires, and thoughts" (p. 47).

4.5.2 *Informants*

This research involved 38 informants from 19 households that adopted biogas technology. These informants included 19 wives and 19 husbands. Prior to the research, the author conducted two days of preliminary intensive observations, which revealed that both wives and husbands played key roles throughout the entire process, from the pre-construction stage to the final use of biogas. This selection aligns with the categorization outlined by the "personal construct" theory, which suggests that direct users should be considered as research informants. The interviews and observations were conducted following McCarthy and Wright's (2004) approach to experience-centered investigation by putting participants' lived experiences and their cultural responses at the center of analysis.

Semi-structured interviews, each lasting 60–90 min, provided the participants with an opportunity to express their genuine experiences and thoughts regarding the use of biogas technology. These discussions covered everything from the initial introduction of the technology to its successful integration into their households and community (Table 4.3).

Besides household informants, the author also involved 5 NGO staffs stationed in Yogyakarta city and Solo Central Java who have served for Biogas Biru Project since 2010. Inquiries in regard with the choice of areas for disseminating biogas, designing of community outreach to communicate with wider potential household users, and their social associations with the users (Table 4.4).

4.5.3 *Participant Observation and Interviews*

Participating in local meetings between biogas users and the biogas extension office greatly enriched my understanding of the strategies used to introduce the technology to local farmers. From the outset, my presence in the field was acknowledged by the users, who recognized me as a researcher since my first visit. This opportunity was

Table 4.3 List of user household interviewed

Name*		Domiciliary	Date of interview and observation
Wife	Husband		
SU	SK	Srihardono Pundong district, Bantul	July–June 2011; July 2012
MW	MJ	Srihardono Pundong district, Bantul	June–July 2011; August 2012
SN	WJ	Srihardono Pundong district, Bantul	June–July 2011; August 2012
PI	PU	Srihardono Pundong district, Bantul	June–July 2012; August 2012
ST	SJ	Srihardono Pundong district, Bantul	June–July 2012; August 2012
TI	TG	Srihardono Pundong district, Bantul	June–July 2012; August 2012
TS	LS	Srihardono Pundong district, Bantul	June–July 2012; August 2012
SY	SG	Srihardono Pundong district, Bantul	June–July 2012; August 2012
SF	SD	Srihardono Pundong district, Bantul	June–July 2012; August 2012
DT	DM	Srihardono Pundong district, Bantul	June–July 2012; August 2012
NT	NG	Srihardono Pundong district, Bantul	June–July 2012; August 2012
MT	MH	Srihardono Pundong district, Bantul	June–July 2012; August 2012
RT	RK	Srihardono Pundong district, Bantul	June–July 2012; August 2012
TR	TW	Srihardono Pundong district, Bantul	June–July 2012; August 2012
SY	SS	Srihardono Pundong district, Bantul	June–July 2012; August 2012
ME	BW	Telaga Adi, Kretek District, Bantul	June–July 2012; August 2012
AH	AG	Sb Mulyo Bambanglipuro district, Bantul	July–August 2012
KI	KT	Sb Mulyo Bambanglipuro district, Bantul	August–September 2012

*Pseudonym

Table 4.4 List of NGO staffs interviewed

Name*	Position	Affiliation	Date of interview	Interview locations
MA	Extension Officer	Biogas Biru	August & September, 2012	Yogyakarta; Solo
WL	Office Manager	Biogas Biru	August & September, 2012	Solo
BA	Quality Assurance Officer	Biogas Biru	August & September, 2012	Solo
TN	CPO Officer	SPTNHPS	August & September, 2012	Yogyakarta
SK	SPTNHPS Officer	SPTNHPS	August & September, 2012	Yogyakarta
YK	SPTNHPS Officer	SPTNHPS	July 2012	Yogyakarta

*Pseudonym

invaluable, as the author participated in local biogas trips to Ngawen, Central Java. During these trips, the author engaged with other biogas users and had the chance to observe and take part in organic fertilizing activities using biogas by-products. This experience significantly enhanced my understanding, as it allowed me to grasp the connection between the technical application of biogas and its social impact on household activities. The method proved to be highly effective in gathering data and deepening my socio-technical insights (Table 4.5).

Participant observation also provided flexibility in exploring the social spaces within households. By "social space," the author refers to the areas and rooms within the home that have been transformed following the adoption of biogas technology. Inspired by Hansen and Winther's (2010) concept of family space, these changes may manifest as modifications to physical interiors, such as the dining room or kitchen, in response to the integration of the new technology. In doing so, the author visited the kitchens of biogas users and photographed them. The users were not hesitant to invite me into their homes; in fact, they felt appreciated and intrigued as the author admired their new interiors. Through this process, the author engaged in warm conversations, during which they shared stories about the physical state of their homes before and after the introduction of biogas technology. By immersing himself in the household and observing the transformation of social spaces, the author was able to better understand the symbolic value of the artifacts. This approach highlights the importance of fieldwork, which can be seen as the art of conversation (Hutchinson, 1996; Wikan, 1999).

Table 4.5 Thematic questions

Framework of domestication	Topics of questions	Research participant	Gathering of data
Appropriation	Socialization of biogas technology, strategies of financing the materials, information and knowledge accessibility, acquisition of material construction	Users, NGO staffs	Semi-structured interviews, participant observation
Objectification	Difficulties in technical knowledge of biogas, Shared responsibility, shifting strategies to migrate from the previous cookstoves, social and cultural preference, motivating values of biogas use	Users	Semi-structured interviews, participant observation
Incorporation	Process of self-habituating, changing in time management and kitchen space, dealing with incidental issues and seeking alternatives	Users	Semi-structured interviews, participant observation
Conversion	The arrival of symbolic meaning after the use of biogas, responds of neighbors, social openness, the social relevance of technology	Users	Semi-structured interviews, participant observation

During the six months of fieldwork, the author lived in a local household, which provided me with firsthand experience of witnessing how biogas was adopted by the families. At the end of each interview, the author always asked the informants to demonstrate their ability to maintain the biogas components, such as the cow dung mixer, the manometer for measuring gas volume, and the biogas stove. This process often helped informants feel more relaxed, encouraging them to share their experiences and recount their early interactions with biogas. The effectiveness of this observation aligns with Darlington and Scott's assertion that observation is a "visible approach to finding out what people do in particular contexts, the routines and interaction patterns of their everyday lives" (2020: 74). In the mornings, the author visited users for two hours to observe how they utilized biogas, documenting my findings in a research diary. Engaging in congenial conversations with family members also granted me valuable insights into their perceptions of this newly adopted technology.

The active re-interpretation of biogas users, shaped by their knowledge and prior experiences with livestock, has led to adaptations in how they use the technology. For example, AWFDFFM reported an extraordinary case where one user mixed chicken quills with cowshed waste, a practice worthy of investigation. It can be inferred that this user had a distinctive cognitive approach, influenced by his experience handling quills. The author visited his home to explore this practice further, realizing that observation does not always need to follow a rigid structure as long as there is a clear purpose. In support of this, Punch emphasizes that an "observational strategy begins with selecting a setting and gaining access to it, then starting the observing" (2005: 180). In addition to formal interviews and observations, informal conversations provided an alternative way to explore issues that might not be captured through structured methods. Spending early mornings at the homes of local leaders and the village secretary after completing interviews was particularly engaging. Over a cup of tea, we enjoyed discussions that allowed me to delve into their aspirations for the biogas project in the village. Additionally, during my morning grocery runs at the local shop for breakfast with my host parents, these congenial conversations with newfound acquaintances further shaped my understanding of local perspectives.

Other sources that contributed to the analysis included documents and archives related to the project. Obtaining permission from the document owners was an essential ethical step, as these materials might contain confidential information or details not intended for public release. In relation to the biogas project, the author received a copy of the contract between AWFDFFM, acting as the construction partner organization (CPO), and HIVOS Southeast Asia, which outlined the project's initial date and the legal obligations of both parties. The organic farming training was conducted under the framework of the Yogyakarta-Central Java Assistance Programme (YCAP) 2009; however, access to the initial proposal for this project was unfortunately denied. Conducting fieldwork on biogas adoption also allowed the author to interact with various local stakeholders, including the village chief and the household biogas (Biogas BIRU) program manager for the Central Java-Yogyakarta office. In this context, establishing formalities helped me gain the trust and credibility necessary to access the relevant data for my research.

In conjunction with the importance of understanding the organizational strategy for biogas dissemination and staffing, the author benefited from the proceedings of the AWFDFFM evaluation workshop (prosiding lokakarya). This proceedings book provided valuable insights into the five-year evaluation of AWFDFFM's work in partnership with global NGOs. It detailed new experiences in organizing biogas promotion to advance their mission and vision in organic farming and sustainable agriculture.

4.5.4 The Interview Situations

Interviews, informal talks, documentary research, and observation were the primary activities conducted in the field for data collection. Various methods are essential in a case study approach, as noted by Crowe et al. (2011), who state that "a case study involves the widest array of data collection as the researcher attempts to build a depth picture of the case." In this fieldwork, interviews were conducted to gather data necessary for the analysis process. A semi-structured interview format was chosen to systematically stimulate informants to respond to questions aligned with the research objectives. Although the interviews were based on prepared questions, the approach was flexible and adapted to the informant's circumstances on that day. The semi-structured interviews were employed to access the context of individuals' behavior, thereby offering researchers a pathway to understand the meaning behind that behavior (Giridharan & Robson, 2011). The author primarily utilized semi-structured interviews with distinct categories of informants, each addressed with tailored interview protocols containing specific questions.

The interview protocols consist of four divisions. The first section contains questions aimed at obtaining basic information about household profiles. The second section includes more detailed inquiries regarding the processes of biogas implementation, household experiences in adopting biogas, and their perceptions of the technology. The third section investigates the involvement of AWFDDFM in the biogas program, focusing on their strategies to reach a larger audience in Bantul Regency. Finally, the fourth section comprises questions directed at Biogas Promotion Officers about their engagement in the biogas program and their collaboration with AWFDDFM.

Firstly, the household profiles were instrumental in identifying the types of biogas users in Sri Hardono village. Specific questions were designed to gather information on various aspects, including the number of family members residing in the household versus those who have migrated, livestock ownership, education levels, age, gender, housing type, the date of biogas construction, household location, and property ownership such as paddy fields. The benefit of collecting this foundational data is to accurately categorize the informants (Biswas, 2011).

The second interview protocol aimed to uncover users' experiences, impacts, and perceptions related to maintaining the household biogas system. The construction process was the initial phase before the biogas was installed, involving several stages.

Specific questions were posed to the biogas users, such as: (i) how they contacted the CPO to submit the pre-construction form, (ii) how they decided on the location of the biogas plant, (iii) the process of preparing or purchasing materials, (iv) arrangements for food and beverages, and (v) the specific time allocated for operating the biogas system within the household, divided into three parts: morning (04:00–09:00), afternoon (11:00–15:00), and evening (17:00–21:00), among others. The informants in this interview included the husband, wife, and other family members involved in the daily operation of the biogas system. The selection of the husband and wife as key informants was based on their roles and responsibilities outlined in the pre-construction contract, which states that both are responsible for handling legal matters. As such, the author engaged both the husband and wife as primary informants for each household. In instances where either was unavailable due to other commitments, the eldest child was often delegated to participate in the interview. This delegation aligns with Javanese cultural norms, where the eldest child holds cultural authority to represent their parents, particularly in community meetings (Jay, 1969; Mulder, 2005).

The third set of interview protocols focuses on the involvement of local non-government organizations in the household biogas project. According to the contract signed with HIVOS Southeast Asia, the AWFDFFM serves as the construction partner organization (CPO). Their responsibilities include overseeing the construction phases and ensuring that users are equipped with the necessary skills to operate the biogas system effectively. As previously mentioned, AWFDFFM's participation in the initial bid to become the CPO was driven by its mission to integrate biogas with its organic farming program. The interviews were conducted with the Secretary, Chairperson, and an officer responsible for managing the "Biogas Rumah" project in Yogyakarta Province. Key questions aimed to explore the following areas: (i) the motivations and organizational goals for becoming the construction partner organization (CPO) for the household biogas project, as well as long-term achievements; (ii) strategies and approaches used to overcome barriers in knowledge transfer to users, especially as this was their first time organizing a biogas project; (iii) how the biogas project aligns with their organizational values; and (iv) the process through which users perceive, receive, and absorb technical knowledge, including any innovations they have developed as a result of this process.

The primary role of the Biogas Promotion Officer (BPO) in the project is to collaborate closely with the construction partner organization (CPO) in promotional efforts. During the socialization stage, the BPO ensures that the targeted community receives accurate information about biogas installation and the specific processes involved. However, their responsibilities extend beyond these activities. A BPO may conduct incidental visits to users' residences to verify whether proper maintenance is being practiced and whether biogas is being used according to the correct procedures. At the local government level, BPOs play a crucial role in enhancing biogas dissemination through strategic communication, partnerships, and identifying political barriers that may hinder promotion efforts. Key questions raised to the BPO included: (i) strategies to engage local governments and highlight the benefits of biogas for the community; (ii) cultural dimensions that both challenge and facilitate biogas adoption; (iii) methods for transferring technical knowledge in a culturally acceptable manner; and (iv) collaborative efforts with AWFDFFM to ensure the success of the biogas project. The author was invited to accompany the BPO on a visit to a biogas user in Sri Hardono, providing an opportunity to observe the dynamics of users' practices in managing biogas in their everyday lives.

4.6 Data Analysis

The personal construct perspective emphasizes that interviews, with a focus on examining direct experiences and the meanings ascribed to them, are central for data analysis (Creswell, 2007; Crowe et al., 2011). The author thus employed a conventional analysis method, which involved "transcription, reducing data into themes through a process of coding and condensing the codes" (Darlington & Scott, 2020). This method of data analysis offers a practical strategy for unraveling the complex and extensive verbal interviews conducted with research participants.

In this research, the analysis of qualitative data including verbal data, observations, meeting logs, and field notes determines the results. Before the interpretation process begins, the analysis covers several sub-processes, including data reduction, data display, and drawing conclusions (Brewer, 2000). The transcription of interview recordings was the first stage. The author transcribed the conversations in the original language of the informants, Javanese. After completing the transcription, the author supplemented it with additional information obtained from my fieldwork diary and meeting logs. Upon completion, the author reviewed the interview transcripts to identify patterns in the data, which enabled me to create descriptive categories and units. Selective efforts were employed to differentiate which data were useful and correlated with the research questions. Coding was then performed based on the aspects the author wanted to examine and discuss in relation to the research questions. This involved sorting the data according to themes, topics, and relevant issues. Different social research theorists highlight that coding and categorizing are integral parts of analysis that shape the interpretation process (Darlington & Scott, 2020; Sandelowski, 1996). Furthermore, interpretation involves reflecting on the data within the theoretical framework. To grasp the *meaning*, Brewer suggests that "interpretation involves attaching meaning and significance to the analysis, explaining the patterns, categories, and relationships" (2000: 104).

4.7 Confidentiality and Inform Consent

In research ethics, it is a moral obligation for the researcher to ensure the security and protection of informants and involved stakeholders. Even though the research topic may not be considered politically sensitive, it is still essential to guarantee that participants remain safe after the research and are not exploited by individuals with political interests. In research ethics, it is a moral obligation for the researcher to ensure the security and protection of informants and involved stakeholders. Even though the research topic may not be considered politically sensitive, it is still essential to guarantee that participants remain safe after the research and are not exploited by individuals with political interests. In doing so, interviews are always initiated with a clear statement offering participants the choice to be involved in the study. The author explicitly mentions the research objectives to ensure participants understand the purpose of the study and their role in it. It is also made clear that their participation is entirely voluntary and without payment, ensuring transparency and ethical compliance. The informants were also asked whether they would allow the author to disclose their names publicly in the research reports. While all of them consented to have their identities revealed, the author took the initiative to de-identify their profiles in this book, applying pseudonyms to protect their privacy. This decision, despite their permission, was driven by a personal ethical responsibility to ensure their anonymity (Scheyvens et al., 2003).

Additionally, confidentiality entails securing all field notes, tapes, and transcripts. All information gathered during the research process is employed solely for the purpose of examination, analysis, and developing research reports. Being a good field worker involves not only adhering to proper methods and procedures but also establishing trust with participants. A trustworthy researcher who shows appreciation for their participants is crucial to maintaining the ethical integrity of the study (Scheyvens et al., 2003). During fieldwork, researchers often engage with informants or respondents who provide valuable data through interviews and other interactions. These informants, having contributed significantly to the research, deserve not only protection but also appreciation for their testimonials and statements. As a researcher, it is essential to adhere to ethical principles that ensure the dignity, privacy, and safety of participants (Scheyvens et al., 2003: 140). Research ethics guide the proper treatment of informants. Two key ethical procedures informed consent and anonymity are central to this practice (Scheyvens et al., 2003).

Informed consent involves ensuring that informants are fully aware of their participation in the research and that their involvement is entirely voluntary, without coercion. Informants must also be aware that they have the right to withdraw from the research at any point if they choose. During the fieldwork with local informants in the village, the author applied this procedure by first asking for their permission to record our conversations. The author clearly explained my status as a research scholar and outlined the purpose of the research, which would ultimately be published online. This transparency helped to build trust with the participants. None of the informants expressed reluctance to be recorded, and though the process felt somewhat formal, it did not create any ethical concerns for me as the researcher. Their understanding and willingness to participate demonstrated that they were cognizant of their role in the research and had voluntarily agreed to share their experiences with biogas technology (Scheyvens et al., 2003: 142).

As a male researcher in these studies, my interactions with informants—particularly housewives who use biogas have generated unique challenges. In many cases, when the author approached women for interviews, they would defer to their husbands, expressing that they lacked sufficient knowledge about biogas technology. This situation presented recurring difficulties during data collection. However, the found a workable solution by positioning himself as a student with limited technical expertise in biogas, genuinely interested in learning how the technology is used daily. By framing the interaction this way, housewives became much more receptive and willing to participate in the interviews about biogas technology.

This dynamic is reflective of the high-context cultural settings often found in the Global South, where societal norms shape communication and gender roles. In such cultures, women may be less inclined to interact with male researchers, particularly in formal or technical discussions, as they are rarely assigned the role of spokesperson for their family or community. This cultural context helps explain why women might hesitate to engage directly in the research process. It reveals the subtle but significant impact that gender roles can have on fieldwork, especially in settings where traditional expectations shape the behavior of individuals during interactions with outsiders.

In essence, my experience highlights the importance of cultural sensitivity and adaptability when conducting research in high-context societies. Understanding the underlying social norms allowed me to navigate the situation more effectively and gain valuable insights from participants who may have initially felt uncomfortable. This notion could help to realize why they in some situations refuse my offer of being interviews in their homes; "Women's freedom in public domain may also be constrained, meaning that they will attend community meeting and hesitantly respond the question".

In fieldwork, effective interpersonal communication hinges on a strong understanding of local knowledge, which is vital for gathering comprehensive and contextually relevant data. For my research, the author conducted interviews with all informants in Javanese, my mother tongue. This proficiency in the local language proved to be immensely advantageous, fostering a deeper acceptance within the community and allowing people to feel more comfortable and open with me. As Rogers (2019) suggests, using the local language enhances cultural understanding and boosts one's ability to comprehend the nuances of local culture. He notes, "knowledge of the local language enables richer and more textured data to be collected and generates greater opportunity to interact and enjoy the company of others."

Beyond communication, the authors' ability to speak Javanese helped the author establish meaningful relationships with staff members of the construction partner organization and some local informants. These connections not only facilitated smoother data collection but also enriched the author cultural sensitivity, allowing me to better appreciate the diversity of ideas and perspectives within the community. Cupples and Kindon (2003) highlight that such relationships can transform a researcher's perspective, helping us view the community as active agents in shaping their own history. This deeper engagement allows researchers to move beyond a detached, observational stance, fostering a more collaborative and empathetic approach to understanding the dynamics at play within the society. By embedding myself within the local context, the author was able to collect more nuanced, culturally informed data, which ultimately led to a more authentic and holistic understanding of the biogas technology and its role in the community. In the final phase of this research, after nearly completing data collection, the author revisited the informants to maintain rapport and share a brief overview of the preliminary findings.

Additionally, my connection with AWDFFDM, the construction partner organization, was not concluded at that stage. The author maintained contact with some staff members, exchanging text messages to stay updated on the progress of biogas adoption in their working area of Bantul Regency. Once the data collection was complete, the author submitted a 10-page report to AWDFFDM in September 2012, detailing the preliminary results. This restitution process proved to be an effective way to gather further insights from the organization, as they shared ideas, views, and opinions regarding the findings. Their feedback was crucial in refining the study and considering future research topics. This approach, in line with Jaeger et al. (2012), serves as a meaningful strategy for validating preliminary findings based on input from local stakeholders. This process not only deepened my understanding of the local setting but also facilitated the development of a more culturally attuned and well-rounded interpretation of the research data. By engaging with AWDFFDM and the local stakeholders throughout this feedback loop, the authored ensured that the research findings resonated with the realities on the ground, while also opening pathways for potential future research collaborations. This participatory approach strengthened the overall quality and relevance of the research.

4.8 Dependability, Conformability, and Reflexivity

Research is often viewed as a legitimate scientific product when it demonstrates validity. As Punch (2005) explains, research can be considered valid if its variables can be measured and trusted by other scholars. In other words, standardized methods for data generation and selection must be both acceptable and testable, ensuring that the same data can be replicated in different settings. This process helps ensure the research is objective. However, while such objectivity is important in quantitative research, it becomes more complex in qualitative studies. For example, a case study, which often stems from qualitative inquiry, prioritizes the depth and uniqueness of a particular phenomenon or community. Striving for strict reliability in this context can reduce the richness of the research, as Balat and Balat (2009) argue, shifting it toward a more positivist approach that might overlook the nuances of human experience. However, in a case study rooted in qualitative inquiry, an emphasis on reliability can lead the research to adopt a positivist perspective (Balat & Balat, 2009), potentially diminishing its interpretive richness. As previously highlighted, a case study is selected specifically because it effectively reveals the particularities and uniqueness of a phenomenon, community, or distinctive situation within a societal group. This approach allows for a deeper understanding of the complexities and nuances of human experiences that may be overlooked in more rigid frameworks.

Confirmability is a key consideration in this study. Robinson (2014) elaborates that confirmability represents objectivity by providing materials for data collection, including raw data (such as documents) and processed data (like interview transcripts and logbooks). Dependability, on the other hand, can be demonstrated by clearly presenting the research process. To address this, we have endeavored to detail the research processes as thoroughly as possible in the methodology chapter and document all relevant materials. Additionally, data triangulation by involving other informants or key actors to verify the accuracy of accounts (Punch, 2005) contributes to supporting dependability. For instance, many users were uncertain about the construction completion date of the biogas plant, which led to some confusion. To resolve this issue, we cross-checked the information against progress reports provided by the Biogas Rumah Office, headquartered in Central Java.

The active involvement and closeness with the users during the research may influence how the author interpreted the verbal data. To mitigate this possibility, critical reflection is essential. As Wong (2007) articulates, critical reflexivity is a self-critical exercise that occurs throughout the research process. During interviews, some biogas users expressed reluctance to respond to my questions, often viewing themselves as less educated. This situation could potentially affect the way they recount their daily experiences with biogas usage. To address this issue, the author emphasized that their experiences are equally important and significant for enhancing my understanding of biogas use, especially since the author had not been directly involved in the construction processes and lack firsthand experience in using the technology In addition, gender may also play a significant role in the research process. As a male researcher interacting primarily with female informants, particularly housewives, the author encountered unique experiences that may have influenced the flow of our interviews. The author's status as a male could create a dynamic that made interactions feel less fluid and potentially impact the quality of their retellings regarding their experiences and perceptions of biogas use.

Despite those challenges, the author made every effort to ensure my informants felt comfortable before the interviews began. Engaging them with non-sensitive questions about their hobbies and activities during their spare time, as well as incorporating light-hearted jokes, helped to alleviate any tension. De Laine (2000) emphasizes that researchers should remain conscious and sensitive to the conditions of their informants, and the author strove to embody this principle throughout my interactions. By applying individual experiences with biogas use, this research aims to investigate how biogas technology navigates the social and cultural contexts of its users, allowing them to integrate it into their daily routines. The results of the fieldwork are primarily ethnographic data that illuminate the social meanings behind the practice and utilization of biogas technology. In this qualitative research, questions of validity and objectivity arise, particularly concerning the scientific justification of the findings. As noted by Punch (2005), the validity of research is established when results are measurable and credible to scholars across various fields. However, this perspective does not hold true for qualitative research, as it may lead to a positivist approach (Jackson et al., 2002). Therefore, we contend that this study is not merely a collection of replicated narratives about biogas technology; rather, it aims to offer readers insights into the dynamics, uniqueness, and specific circumstances that shape how biogas is used, appropriated, and routinized in rural Indonesia.

Throughout the writing process and prior to concluding my time in the field sites, I engaged in critical reflection regarding my role as a researcher. This practice is essential, particularly when it comes to distinguishing between the personal and professional realms during participant observation, as it can lead to potential biases in the findings. Wong (2007) emphasizes that self-reflection is a valuable exercise that occurs throughout the research process, helping to determine what is worth unveiling and what may need to be set aside.

References

Ayuwandani, Y., & Tondok, M. S. (2023). Psychological aspects in the "Nyumbang" tradition of Javanese society. *Santhet: Jurnal Sejarah, Pendidikan, Dan Humaniora, 8*(1), 178–187. https://ejournal.unibabwi.ac.id/index.php/santhet/article/view/3326

Bakardjieva, M. (2006). The consumption junction revisited. *Computers, Phones, and the Internet: Domesticating Information Technology, 2*(1), 97–98.

Balat, M., & Balat, H. (2009). Biogas as a renewable energy source—A review. *Energy Sources, Part a: Recovery, Utilization, and Environmental Effects, 31*(14), 1280–1293. https://doi.org/10.1080/15567030802089565

BAPPEDA Bantul. (2009). *Regional village profiles*. BAPPEDA.

Berker, T., Hartmann, M., & Punie, Y. (2005). *Domestication of media and technology*. McGraw-Hill Education.

Biswas, W. K. (2011). Application of renewable energy to provide safe water from deep tubewells in rural Bangladesh. *Energy for Sustainable Development, 15*(1), 55–60. https://doi.org/10.1016/j.esd.2010.10.003

Brewer, P. J. (2000). *From fireplace to cookstove: Technology and the domestic ideal in America*. Syracuse University Press.

Creswell, J. W. (2005). *Educational research: Planning, conducting, and evaluating quantitative and qualitative research*. Pearson.

Creswell, J. W. (2007). *Research design: Qualitative, quantitative, and mixed methods approaches* (2nd ed.). Sage.

Crowe, S., Cresswell, K., Robertson, A., Huby, G., Avery, A., & Sheikh, A. (2011). The case study approach. *BMC Medical Research Methodology, 11*(1), 1–9.

Cupples, J., & Kindon, S. (2003). Far from being "home alone": The dynamics of accompanied fieldwork. *Singapore Journal of Tropical Geography, 24*(2), 211–228. https://doi.org/10.1111/1467-9493.00153

Darlington, Y., & Scott, D. (2020). *Qualitative research in practice stories from the field*. Routledge.

De Laine, M. (2000). *Fieldwork, participation and practice: Ethics and dilemmas in qualitative research*. Sage.

DeWalt, K. M., & DeWalt, B. R. (2002). *Participant observation: A guide for fieldworkers*. Altamira Press.

Giridharan, B., & Robson, A. (2011). *Identifying gaps in academic writing of ESL students*. In Enhancing Learning: Teaching and Learning Conference 2011, Curtin University Sarawak.

Hall, D. (2003). The international political ecology of industrial shrimp aquacultural and industrial plantation forestry in SouthEast Asia. *Journal of Southeast Asian Studies, 34*(2), 251–264. https://doi.org/10.1017/s0022463403000249

Hutchinson, G. (1996). Threats to underwater cultural heritage. The problems of unprotected archaeological and historic sites, wrecks and objects found at sea. *Marine Policy, 20*(4), 287–290. https://doi.org/10.1016/0308-597x(96)00017-6

Jackson, M. H., Poole, M. S., & Kuhn, T. (2002). The social construction of technology in studies of the workplace. In L. A. Lievrouw & S. Livingstone (Eds.), *Handbook of new media: Social shaping and consequences of ICTs* (pp. 236–253). Sage.

Jaeger, P. T., Bertot, J. C., Thompson, K. M., Katz, S. M., & DeCoster, E. J. (2012). The intersection of public policy and public access: Digital divides, digital literacy, digital inclusion, and public libraries. *Public Library Quarterly, 31*(1), 1–20. https://doi.org/10.1080/01616846.2012.654728

Jay, R. R. (1969). *Javanese villagers: Social relations in rural Modjokuto*. Cambridge University Press.

Lestari, S., Adiyanti, G., & Walgito, B. (2013). The concept of harmony in Javanese society. *ANIMA Indonesian Psychological Journal, 29*(1), 24–37.

McCarthy, J., & Wright, P. (2004). Technology as experience. *Interactions, 11*(5), 42. https://doi.org/10.1145/1015530.1015549

Mulder, N. (2005). *Inside Indonesian Society: Cultural change in Java*. Kanisius.

Nuradita, R. D., & Subiyanto, A. (2024). Etnografi Komunikasi Ritual Brokohan Sapi di Desa Sendangwungu Kabupaten Blora. *Jurnal Onoma Pendidikan Bahasa Dan Sastra, 10*(4), 3825–3834. https://doi.org/10.30605/onoma.v10i4.4406

Østern, T. P., Jusslin, S., Nødtvedt Knudsen, K., Maapalo, P., & Bjørkøy, I. (2023). A performative paradigm for post-qualitative inquiry. *Qualitative Research, 23*(2), 272–289. https://doi.org/10.1177/14687941211027444

Punch, K. F. (2005). Introduction to social research—Quantitative & qualitative approachesial research. *Forum Qualitative Sozialforschung/Forum: Qualitative Social Research, 7*(2).

Ramachandra, T. V. (2008). Geographical information system approach for regional biogas potential assessment. *Research Journal of Environmental Sciences, 2*(3), 170–184. https://www.cabidigitallibrary.org/doi/full/10.5555/20083259740

Robinson, O. C. (2014). Sampling in interview-based qualitative research: A theoretical and practical guide. *Qualitative Research in Psychology, 11*(1), 25–41. https://doi.org/10.1080/14780887.2013.801543

Rogers, T. L. (2019). Reciprocity and visual methods in cross-cultural research with Cambodian schoolgirls. *Children's Geographies, 18*(4), 1–15. https://doi.org/10.1080/14733285.2019.1668910

Sandelowski, M. (1996). One is the liveliest number: The case orientation of qualitative research. *Research in Nursing & Health, 19*(6), 525–529. https://doi.org/10.1002/(SICI)1098-240X(199612)19:6<525::AID-NUR8>3.0.CO;2-Q

Scheyvens, R., Wild, K., & Overton, J. (2003). International students pursuing postgraduate study in geography: Impediments to their learning experiences. *Journal of Geography in Higher Education, 27*(3), 309–323. https://doi.org/10.1080/0309826032000145070

Sullivan, S., & Brockington, D. (2003). Qualitative research. In *Development fieldwork* (pp. 57–74). Sage.

Susilo, S., & Syato, I. (2016). Common identity framework of cultural knowledge and practices of Javanese Islam. *Indonesian Journal of Islam and Muslim Societies, 6*(2), 161. https://doi.org/10.18326/ijims.v6i2.161-184

Wikan, U. (1999). Culture: A new concept of race. *Social Anthropology, 7*(1), 57–64. https://doi.org/10.1017/s096402829900004x

Winther, T. (2010). *The impact of electricity: Development, desire and dillemas*. Berghahn.

Wong, S. (2007). Partner or pariah?: Building social capital with clan associations in Hong Kong. *Chinese Sociology & Anthropology, 40*(1), 54–71.

Woods, C. T., & Davids, K. (2023). Thinking through making and doing: Sport science as an art of inquiry. *Sport, Education and Society, 28*(5), 579–593. https://doi.org/10.1080/13573322.2022.2054792

Chapter 5
Appropriating and Objectifying Biogas

Abstract When biogas emerges in the market as an alternative source of household energy, potential customers often have access only to basic information and expertise. This limited understanding can hinder their ability to fully grasp how the technology can be employed effectively. In this chapter, we will explore how biogas is appropriated within domestic spaces and how it inspires users to make innovative modifications to their routines and behaviors. Key aspects discussed include the educational attainment of biogas user households, the valuation of construction materials, the dynamics of biogas usage in everyday life, villagers' exposure to biogas technology, family support for purchasing construction materials, kinship support, and the interplay of gender relations in biogas use. As biogas is recognized in the market as an alternative household technology, the knowledge and information available to potential users are critical to enabling a comprehensive understanding of its application. This chapter delves into the ways biogas is integrated into household life and how it serves as a catalyst for changing habits and customs.

Keywords Appropriation · Biogas · Users · Non-users · Family · Gender

5.1 Structure of the Family

This study focuses on the transformative life cycle of biogas, examining how users' households alter their lives after adopting this technology. The households included in the research were selected based on specific criteria: each installation must have been in place for at least three months, and the users must have actively utilized it for their household needs and farming. Given the geographical constraints of the research, which involved households spread across a challenging terrain, the author selected households within the same village and its neighboring areas. In identifying suitable households for the study, the author followed the advice of AWFDFFM, an organization that is well-acquainted with the local context and can facilitate access within

M. Alam, *The Social Life of Biogas*, https://doi.org/10.1007/978-981-96-3300-5_5

limited timeframes. This purposive sampling approach, supported by the organization involved in the construction of biogas systems, proved beneficial for achieving the research objectives (Mazunda & Shively, 2015).

Instead of solely relying on interviews and observations, as suggested in the personal construct method, the author also summarized survey data to provide readers with insights into the social characteristics of the households involved in the study. The survey encompasses demographic information for each household, including the number of family members, occupations, number of livestock, educational backgrounds, and details about the biogas installation. Since my informants are husbands and wives who have direct experience utilizing the biogas system, the demographic data is intentionally designed to profile them. Additionally, the author included information about the children living in the household to provide a more comprehensive understanding of the family dynamics.

5.1.1 Structure of Household Users of Biogas

Instead of solely relying on interviews and observations, as suggested in the personal construct method, the author also summarized survey data to provide readers with insights into the social characteristics of the households involved in the study. The survey encompasses demographic information for each household, including the number of family members, occupations, number of livestock, educational backgrounds, and details about the biogas installation. Since my informants are husbands and wives who have direct experience utilizing the biogas system, the demographic data is intentionally designed to profile them. Hence, the author included information about the children living in the household to provide a more comprehensive understanding of the family dynamics.

Consumption timing significantly influences the volume of biogas available, serving as a critical factor in the socio-technical dynamics within the family. Mornings can be particularly hectic, especially when children are getting ready for school and housewives prepare meals for their husbands. During this peak time, the household typically engages in boiling water for drinking, cooking rice, and preparing other dishes, all of which contribute to a higher demand for biogas. Evenings also represent a peak period when families gather to enjoy dinner together. In contrast, biogas use tends to diminish in the afternoons, as many family members are away at work or school.

Interviews reveal that all households are aware of biogas limitations and therefore exercise careful control over their cooking activities. Additionally, since these households also care for cows, they utilize the morning hours to prepare rice chaff for their livestock. This process takes over ten minutes, during which they rely on LPG cookstoves. Consequently, it is evident that non-biogas cookstoves remain essential for fulfilling daily energy needs, especially for animal feed preparation.

5.2 Educational Attainment of User Households

When it comes to education, technological acceptance often requires users to possess sufficient know-how, including at least basic reading skills. However, according to the survey conducted for this study, many users struggle with reading even simple manuals, such as those guiding the use of the manometer or instructing how to switch on the cookstoves. Literacy skills, which can influence the ability to use technology effectively, are often shaped by the level of education the informants have received. As illustrated in Fig. 5.1, senior high school is the most commonly attended educational institution among the participants. Notably, there is no significant difference in educational attainment between husbands and wives at this level. A university degree is the second most attended institution, although only a few individuals pursued this level of education due to their remote location, far from the city. According to the interviews, users revealed that education has little impact on accessing technical knowledge related to biogas, particularly in reading the manometer or following the guidelines for regular biodigester feeding. Instead, experience and persistent habituation in using biogas are the key factors for understanding its technical aspects. This aligns with Joshi, Chelliah, and Ramanathan's (2015) perspective on grassroots innovation, where local capacities and community initiatives are more crucial than formal education, with mutual assistance naturally emerging in these contexts:

> Grassroot innovation (GI) a mind-set-driven phenomenon where acumen, capability and belief system of the individual innovator plays an important role, a solution science that enhances livelihood or existing process or product for the greater-good of the community and driven by trust and reciprocity. (p. 39)

From the interviews, all informants confirmed that the knowledge gap is bridged through intensive training organized by AWFDFFM, and they are encouraged to raise questions at any time via mobile contact. An overview of the overall education profile of the biogas users is presented in Fig. 5.1.

5.3 Comparing the Valuation of Construction Materials

Despite the fact that cows are a primary requirement for biogas adoption in this program, very few users do not own them. As shown in Table 5.1, only two users lack cows, yet they experience no difficulties with daily feeding of the biogas digester. They source cow dung from neighboring households that have cows but do not adopt biogas due to a lack of available land. For these two users, the absence of cows is not a significant issue, as they can find alternatives. One user even utilizes pig dung, as he operates a pig breeding business at home with 12 pigs. After several years, he found that the pig dung became a significant hygienic concern for his neighbors due to the unpleasant odor and the difficulties in disposing of it. Therefore, installing a biogas system was a far better solution for converting this animal waste. As elaborated in the preceding chapter of the introduction, users are required to contribute to

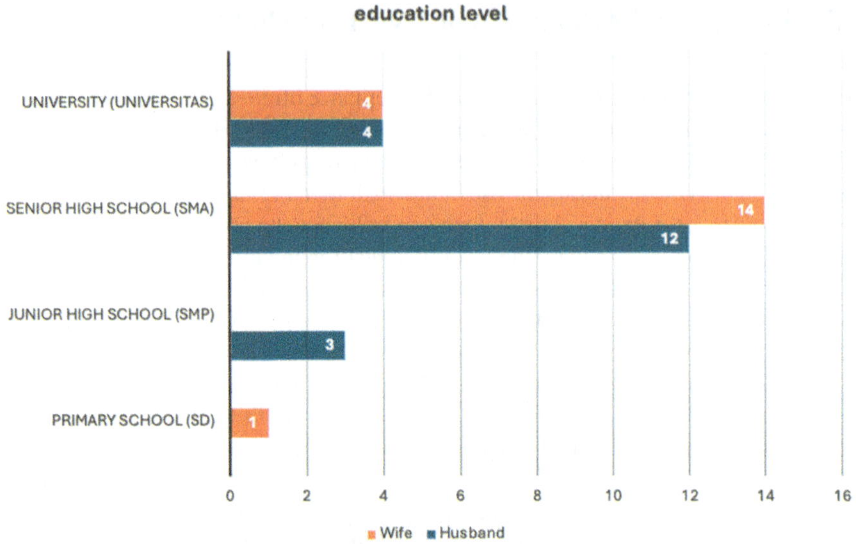

Fig. 5.1 Bar chart comparing education levels of husbands and wives. Levels include primary school, junior high school, senior high school, and university. The chart shows that more husbands than wives have only primary education, while both have equal university levels. Most husbands have senior high school education (*Source* author)

the construction materials by either providing financial support to the construction project organization (CPO) or supplying the materials themselves. Self-provisioning of materials offers numerous benefits, including cost savings, the freedom to choose affordable materials, and the ability to customize the necessary materials to meet their specific needs.

Table 5.2 illustrates the variation in construction material costs from one household to another. Households that chose to provide their own construction materials incurred lower expenses overall. As reflected in the table, these differences in material valuation likely stem from the specific choices each household made. Many informants shared during interviews that they reused leftover materials from post-earthquake reconstruction, such as bamboo, bricks, and even cement.

However, geographical conditions also played a role in affecting the construction process. For instance, during March 2010, heavy rainfall in the area significantly impacted building efforts. BW, one of the informants, reported higher expenses due to the need to purchase additional materials like tarpaulins and polyethylene plastic to reinforce the ground layer, as water was resurfacing from below.

According to interviews with informants, cow ownership is closely linked to the cultural status of villagers. The cultural notion of "*Raja Kaya*" which translates to "the rich man" suggests that owning cows symbolizes high social status among rural households. By possessing such livestock, individuals are often regarded by their peers as more affluent compared to others in the village. In times of economic

Table 5.1 Family members of the interviewed household

No	Name of household head	Children living at home		Total number of family members at home	Occupation	
		Male (Age)	Female (Age)		Husband	Wife
1	SK	–	2	4	Masseur	Migrant worker
2	MJ	3	–	4	Farmer	Housewife
3	WJ	1	1	5	Hospital officer	Organic paddy farmer
4	PU	1	1	4	Farmer	Farmer
5	SJ	2	–	4	security Guard	Credit officer
6	TG	1	1	4	Farmer	Teacher
7	LS	–	–	2	Farmer	Farmer
8	SL	2	2	6	Farmer	Farmer
9	SG	2	1	5	Farmer	Merchant
10	SD	1	2	5	Government officer	Housewife
11	DM	2	1	5	Farmer	Housewife
12	NG	1	2	2	Chicken trader	Nurse
13	MH	–	–	4	Organic paddy farmer	Farmer
14	RK	1	1	5	Farmer	Housewife
15	TW	3	–	5	Police	Housewife
16	SS	3	–	5	Police	Housewife
17	BW	1	1	4	Cow broker	Herbal practitioner
18	AG	–	2	4	Civil police	Housewife
19	KT	3	–	5	Farmer	Housewife

hardship, cows are frequently sold to address immediate financial needs. Informants shared that when their children were preparing for wedding ceremonies, they had offered their cows to neighbors for sale at reasonable prices to alleviate financial constraints. Similarly, parents often find themselves in situations where they need to pay school fees before the start of a new academic year. Given that schools charge substantial fees, households typically resort to selling their cows, especially between June and August, when most public and private schools begin new enrollment. This situation partially supports the author's assumption regarding the low rate of biogas adoption in rural areas; user households often prioritize spending their money on their children's education rather than investing in biogas construction In an interview with the CPO (Community Project Officer) involved in post-earthquake recovery

Table 5.2 Valuation of construction materials

Name of household head	No. of livestock	Completion date of construction	Value of construction material (in USD)
SK	Not available	08.08.2010	160
MJ	Not available	09.11.2010	188
WJ	2	02.11.2011	172
PU	2	09.11.2010	190
SJ	2	16.11.2010	180
TG	1	02.02.2011	167
LS	3	12.10.2010	180
SL	2	09.03.2011	177
SG	4	29.10.2010	167
SD	1	11.11.2010	190
DM	2	21.07.2010	Not available
NG	None	19.03.2010	150
MH	6	27.02.2011	200
RK	1	25.04.2011	170
TW	3	09.03.2011	250
SS	2	09.03.2011	280
BW	30	19.03.2011	260
AG	2	19.08.2010	200
KT	2	24.08.2010	230

programs and organic farming initiatives, it was revealed that most rural households in Bantul continue to face financial constraints, and their economies have not fully recovered since the earthquake in 2006. This ongoing economic challenge significantly hampers efforts to reach potential adopters in the area. People tend to prioritize financing their basic needs before considering additional expenses, which can be substantial. During our conversation, it became clear that biogas adopters in Bantul are primarily those who have sufficiently recovered from financial hardships and achieved economic stability. The presence of fallow lands in Bantul, while seemingly conducive to biogas adoption, does not necessarily guarantee a higher adoption rate, contrary to predictions made by SNV in their 2009 assessment.

Moreover, statements from sub-village chiefs provide contrary insights. They reported that, to date, 80% of local rural lands in Sri Hardono village are traditionally inherited without legal documentation, leading to ownership disputes that remain unresolved. The elaboration above sheds light on the complexities surrounding the low adoption rate of biogas technology in the area. Notably, the abundant availability of land and livestock are not the primary factors that promote the acceptance of biogas technology.

5.4 Appropriating Biogas

As explained in Chapter 3, appropriation is the stage at which biogas begins to be acquired through a variety of social activities such as infrastructure development, feedstock procurement, user training and workshops, and initial monitoring.

5.4.1 *The Villagers' Exposure to Biogas Technology*

Before users bring biogas technology into their homes, their adoption practices are influenced by information and knowledge obtained from various sources. The ways in which users accept and adopt such mundane information are worth exploring. This section unpacks the various sources of information to which users were exposed and, more importantly, examines how these sources influence their intention to adopt biogas technology. In acquiring biogas technology, many scholars have investigated several key factors that contribute to its acceptance within the community. These factors include a comprehensive understanding of how the technology works, effective awareness generation campaigns, the dissemination of information, proper identification of targeted users, and the mobilization of resources from NGOs (Mshandete & Parawira, 2009; Pal, 2003; Parawira, 2009). Each country presents a unique context that can facilitate the adoption process. In this regard, the author will present findings on how knowledge and information were accessed, which ultimately proved beneficial to the users. To facilitate the adoption of biogas technology, a "demonstration plot" was established in Sri Hardono as a designated learning space for educating construction laborers (Lewanika & Mulenga, 1996; Pal, 2003). This village was selected because 90% of the locals participated in the Yogyakarta Community Action Program 2007, organized by AWFDFFM to enhance local livelihoods through organic farming.

Furthermore, the training workshop for construction laborers received significant coverage from national media at that time (Interview with TN). It is not surprising that the largest number of biogas adopters in Bantul resides in Sri Hardono village, as interviewees frequently recognized the village's name. The dissemination of information regarding biogas application was carried out collectively by AWFDFFM, the construction partner organization, which aimed to expose community members to the sustainable benefits of biogas installations for households. Given that the village consists of 17 sub-villages and considering the limited time available, the introductory meeting likely influenced local perceptions regarding biogas adoption. Although the event was attended by sub-village heads and household leaders, AWFDFFM anticipated that these attendees would relay the information to all interested local residents. However, when the author inquired about the lack of a significant increase in biogas users in Sri Hardono, AWFDFFM explained that the project prioritized reaching quality biogas users over merely increasing numbers, largely due to limited manpower available for the initiative. Information about biogas installation in the village is crucial, as it serves as the foundational knowledge necessary for potential users to understand the technology. In the early phase of its introduction to Bantul, biogas was first presented by AWFDFFM through an introductory meeting that included approximately 20 heads of households. However, users accessed information through various mediums and channels.

In June 2010, AWFDFFM, in collaboration with BIRU program officers from HIVOS Southeast Asia, conducted an introductory meeting at the community hall in Jonggrangan sub-village, Sri Hardono. During this meeting, AWFDFFM and BIRU program officers outlined their respective roles and responsibilities. The BIRU program officers provided a general overview of the biogas program, which would run until December 2012, highlighting the benefits of biogas technology for enhancing family economies. Meanwhile, AWFDFFM delved into the specifics of the construction process, including the materials required. AWFDFFM began their presentation by discussing the different models of biogas that households could choose from, tailoring the options to their family needs. They emphasized the potential to reuse leftover construction materials available at home. As noted by TN, potential users may feel discouraged if they are initially confronted with the total cost, as this can create a sense of financial burden. There were a total of 30 attendees at the meeting, but not all registered as biogas users. This hesitation stemmed from various considerations, including concerns about financial sources, alignment with their interests in using alternative energy, the technical requirements for installation, the suitability of land for placing the biogas plant, and potential risks. This situation raised concerns and prompted the author to conduct further investigation.

Among the attendees, 17 were sub-village chiefs, while the remaining participants were local residents interested in learning about biogas technology. Most attendees were men with farming backgrounds who owned at least one cow. At the end of the meeting, those who expressed a strong interest in installing biogas technology registered their family information. This included the number of cows they owned, the locations of their homes, the available land area for biogas plant construction, the distance from the kitchen to the designated land, the distance to the cowshed, the number of family members living at home, the cooking stoves they currently used, and any other construction materials they could provide, such as cement, coral stone, bricks, and sand. Those who registered their names were invited by AWFDFFM for a preliminary survey.

Information technology has also provided users with the advantage of obtaining a complete overview of the technical design process and construction of biogas technology. SS, one of the biogas users, actively uses the internet for daily administrative tasks at his office. While the internet helped satisfy his curiosity about biogas, it also had limitations, particularly for users without sufficient experience in contacting potential construction agencies. This highlights that technology has its limits in translating information into reality, especially when prospective users desire a physical overview of the biogas installation. SS, a policeman living in the Karangasem sub-village of Pundong, Bantul, had long harbored a desire to install a biogas system. After realizing that the cattle dung at his home was excessive, he sought alternatives to convert it into more productive resources. As an active seeker of biogas information, he dedicated much of his lunch break to researching the best biogas plant designs. Before deciding to install a biogas reactor, SS spent over three months exploring biogas information online. His daily internet use at the office allowed him to access various biogas models that suited his needs. He expressed dissatisfaction with two previous biogas reactors from a government-funded project, which had failed and were left without proper maintenance. This experience motivated him to find a more reliable and credible construction organization that could ensure a proper installation and guarantee the technology's durability. To ensure that his preferred biogas design was both sufficient and reliable, SS collected information on various biogas reactors, enabling him to distinguish favorable models and make informed comparisons. During the interview, the author was impressed by SS's remarkable ability to draft the design of the biogas plant he envisioned. He was well-informed about the details and operating system of biogas technology, which he had researched online, and he compared this knowledge with the biogas plants funded by the government. For the preparation of his biogas plant, he meticulously designed its placement in relation to the kitchen, including the necessary length of pipelines to connect the gas output to the cooking area. During his exploration of the desired biogas model, SS indicated that he had allocated a budget of IDR 3,000,000 (approximately USD 300) for the project.

The information about biogas technology often comes from individuals who have benefited from it, and SS's firsthand experience enabled him to share valuable insights with fellow residents. SS and TW, both police officers, exemplified the role of trust and social networking in the acquisition of biogas technology. During a regional police gathering in Bantul regency, SS met TW, who had already constructed a biogas system at his home. This chance encounter provided SS with a valuable opportunity to learn more about biogas and identify potential construction organizations that could support his initiative. Following this meeting, TW arranged more in-depth discussions with SS to explain the differences between various biogas installations compared to models adopted by other users. This exchange proved significant, as it reshaped SS's understanding of the risks and potential consequences associated with biogas technology. Information gained from someone with a similar perspective can effectively reduce or alter the perceived risks for users. SS's reflections on this interaction reinforce the importance of social connections in the adoption process:

SS: "When I receive information from friends who work in the same field, I feel more comfortable, and most importantly, I trust the credibility of the information they provide. It's highly unlikely that close friends would deceive or lie to us—they would give us honest advice, especially when it comes to decisions like what we want to install.

MA: "But then, do you think friends who have adopted the technology can reassure you by claiming that it is not risky?"

SS: "That is definitely true. In the case of biogas, I think his preference was helpful and assured me that it could be categorized as a zero-risk cooking appliance compared to LPG. The initial information I found on the internet about biogas was beneficial because it provided a technical illustration, but I knew it couldn't be completely reliable and might not be entirely accurate. Therefore, checking the information with friends adds significant value to ensure that what I read online is correct. Afterwards, I asked SS to help me contact TN (the AWFDFFM officer) for further construction procedures."

In addition to the experiences mentioned above, the successful dissemination of information among biogas users who did not attend the introductory meeting relied heavily on word of mouth. In this process, the agency played a crucial role. Since AWFFDFM did not have a specific budget allocated for extensive introductory meetings, the importance of social capital in reaching potential users became paramount. To facilitate the flow of information, SK, who had a close relationship with the village officers and the village chief, was informally appointed as an agent responsible for passing on the information to the villagers. SK's informal appointment stemmed from his active engagement in organic farming and his participation in mason training in June 2010.

According to a personal interview with SK, his ability to effectively communicate information about biogas technology to the villagers was rooted in his deep understanding of the local culture in Sri Hardono village and his strong connections with other residents. Figure 5.2 illustrates the interactions among biogas users before the construction processes took place. The arrows indicate the direction of information flow, while the double-dotted arrows mark the interactive processes that developed between users.

In Fig. 5.2, the author outlines how SK served as the agent sharing information with three different users at that time. The interaction first occurred between SK and SJ, who were both in the same sub-village. There was a fixed agreement between SK and AWFDDFM for him to act as an agent to disseminate information about biogas, with the goal of increasing the number of adopters. SK's profile as a loyal individual, along with his ability to approach the village government, made him an appealing choice for AWFDDFM as an informal agent. SK acknowledged that if he could identify one potential user interested in installing biogas, he would receive a certain amount of money from AWFDDFM. Although the specifics of the financial incentive were unclear, it likely motivated SK to spread the information and reach more users. Despite lacking prior technical capacity, SK found it challenging to gauge the interest of other local residents, including SJ. SJ, who worked as a security guard to earn extra income, was motivated to run a small food court at home, a plan he expressed during a social meeting convened by the village chief of Sri Hardono. SK's active participation in village social meetings provided him with greater opportunities to meet potential users. This included persuading TG to install a biogas reactor at home. SK's role extended beyond merely disseminating information about the advantages of biogas; he was also keen to facilitate communication with AWFDDFM as the construction partner organization. TN, acting as a representative of AWFDDFM, appreciated SK's assistance during this process.

At this stage, it became evident that informality among villagers was a crucial factor in accelerating the dissemination of information. The networking channel of

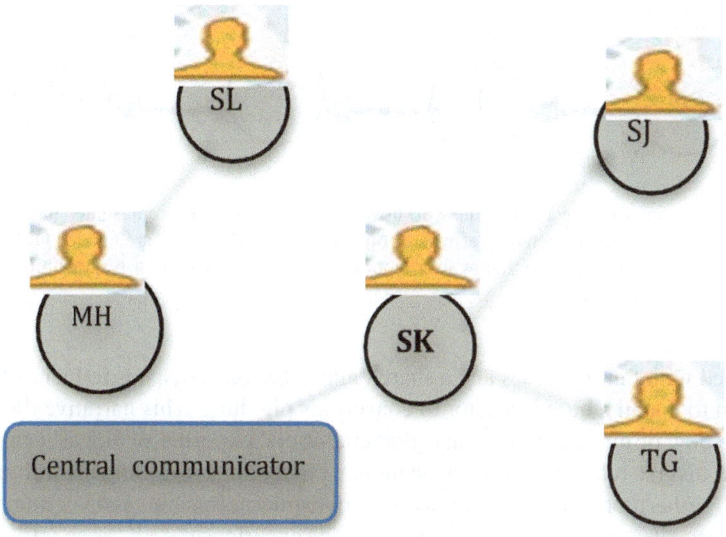

Fig. 5.2 Diagram showing a network of communication between five individuals, each represented by an icon with a silhouette and initials. The icons are connected by lines to a central rectangle labeled "Central communicator." The individuals are labeled as SL, SJ, MH, SK, and TG. The lines indicate communication paths between the central communicator and each individual (*Source* Author)

SK was extensive—not limited to just the Jonggrangan sub-village. As a freelance administrative assistant for the Sri Hardono village office, he had ample opportunities to visit other local residents. One day, during a visit to Gulon sub-village to assist with a demographic survey, he coincidentally met SL, the owner of a 1-hectare palm oil plantation in South Sumatra. Their meetings became more frequent as SL expressed interest in installing a biogas reactor. SL's cattle shed was located just 1 m from a small street connecting Gulon sub-village with Jonggrangan sub-village. During heavy downpours, the dung mixed with rainwater created a strong odor that was unpleasant for anyone passing by. In December 2010, SL received complaints from his neighbors about the smell. Faced with this issue, he felt compelled to find a solution to reduce the odor—biogas seemed like a viable option. He worried that the complaints from neighbors might escalate and lead to greater hostility toward him if he did not address the problem. Additionally, he was distressed by gossip circulating among the neighbors, suggesting that he was an incapable businessman who could not afford to manage his cow dung properly.

On the other hand, SL is the nephew of MH, who lives in Kembangkerep sub-village, about 1 km from SL's home. Once a week, MH visits SL to stay updated on the progress of the plantation. Their close kinship fosters a strong trust between them, which remains intact despite their differing religions. MH's organic rice business has been rapidly growing, and he often seeks SL's assistance in promoting his products or introducing them to prospective customers in Yogyakarta city. This existing trust,

Fig. 5.3 Diagram showing a sequential workflow with four stages, each represented by a gray oval containing initials and connected by arrows. From left to right, the stages are labeled DM, SL, SS, and BW. Above each stage is a yellow user icon, indicating involvement or responsibility at each step (*Source* author)

combined with family values and a shared interest in business, has influenced MH's decision to install a biogas reactor to convert his pig dung. This narrative illustrates how the flow of information impacts potential users' exposure to biogas technology and how diverse relationships reshape mutual perceptions regarding the information shared. Furthermore, the information flow is dynamic and encompasses certain values that enhance its comprehensibility for users. The experience of SK, as an information agent, highlights the strength of interpersonal networks within the village. Similarly, DM, the head of Tangkil sub-village, also played a role in disseminating biogas information among potential users. I will illustrate his role in Fig. 5.3 to depict the linear flow of information.

Based on Fig. 5.3, DM serves as the starting point for the flow of information. It is clear that he successfully persuaded TW to adopt biogas technology for his household. Since DM's land was designated as a demonstration plot, he took pride in promoting the benefits of biogas to those close to him, including friends. TW, in particular, was one of DM's most frequently visited friends, as they both attended the same junior high school 15 years ago. Their enduring friendship is maintained through regular visits, occurring twice a week. This frequency of interaction enhances trust, allowing reliable information to be exchanged between them.

The network chain depicted in Fig. 5.3 also highlights that users who have successfully installed biogas systems can serve as valuable resources for potential adopters they encounter, including TN, the representative of AWFDFFM. The limited capacity of AWFDFFM to disseminate information about biogas throughout the village can, to a large extent, be addressed through such alternative information networks. The underlying distrust in technological promotion, as indicated in the previous statements, underscores that trust is a critical experiential dimension in the adoption of technological artifacts.

5.4.2 *Prior Experience Shaping Preference*

Users' exposure to biogas technology varies significantly, and prior experiences with different types of biogas systems can impact their decisions in later stages. In

2009, a biogas experiment project was organized in the Kembangkerep sub-village by university stakeholders, which influenced users' understanding of biogas technology models. In this context, RK and MH both had unsatisfactory experiences with government- and university-led biogas projects. To illustrate this, the author would like to share RK's story about his grandmother's experience with a newly installed biogas system at her home.

RK's grandmother, who lived in a different village, was among the recipients of a biogas project implemented by a student community program organized by a local public university in Yogyakarta. With a look of frustration, RK recounted how the biogas plant at his grandmother's home broke down just a year after its installation. This issue arose because his grandmother was not provided with adequate instructions on how to maintain the biogas plant, the cookstove, or how to feed it with manure. RK's primary disappointment stemmed not from the technology itself but from the university students' failure to properly convey the necessary maintenance procedures.

> I know they might be just 'hit and run'. Such projects just included us for experimentation," he expressed with anger. This unfortunate experience taught him a valuable lesson about selecting a reliable construction agency, as he wanted to avoid a similar situation in the future. His trust in AWFDDFM for the biogas construction at his own home was reinforced by this past experience. He confidently stated, "AWFDDFM is trustworthy because they offer a three-year service guarantee, and I can consult the local mason if my biogas system has low output.

MH, a resident of Kembangkerep sub-village, recounted a second experience with a similar project conducted by a private university in Yogyakarta in the year 2000. This particular biogas technology model featured a floating dome designed to serve as a gas container. To operate the biogas plant effectively, the dome needed to be pressed down to facilitate gas output. Since the gas was intended for collective use by local farmers, the university experts installed only one plant in the sub-village. Initially, the villagers, including MH, were enthusiastic about the biogas project. However, after several trials, they found the floating dome model to be overly strenuous and complicated, ultimately proving unhelpful for their agricultural practices. Consequently, their interest in this model waned. This experience made MH more discerning regarding the adaptability of biogas applications that might be promoted by other experts in the future. The insights he gained from the trials with the floating dome biogas system equipped him with valuable knowledge, leading him to opt for the BIRU biogas model. He noted that one of the appealing features of the BIRU system is the steel-made inlet basin, which simplifies the process of mixing dung with water, making it more convenient and enjoyable.

These two stories highlight that rational considerations in choosing a biogas reactor are often informed by prior experiences with different models or types of plants. The author interprets that users' previous encounters with biogas systems cannot be overlooked, as these cognitive experiences shape their preferences and guide them toward adopting technologies that align with their needs and circumstances.

5.4.3 Financing Biogas Technology

Financing the construction of biogas systems can be a significant concern for households before the installation process begins. This section explores how families navigate these financial challenges and compares the experiences of those who invest in their biogas systems with those who do not. Given that the initial feeding of the system is a strenuous activity, users encounter this stage differently. In this study, one household was provided a biogas plant at no cost. It is noteworthy to present this case because the author discovered that this user operated the biogas system improperly, unlike others who invested their own money into the construction materials.

This household's land area was designated as a demonstration plot for mason training and served as a center for local biogas exhibitions. The appointment was granted due to the landholder being an open-minded and cooperative resident of the Tangkil sub-village, who had expressed interest to AWFDFFM before the biogas installation. Despite owning cows, he only fed the reactor every two weeks, citing that his work on the farm and administrative tasks in the village consumed most of his time. He remarked that the biogas stove appeared very corrosive, and the outer part of the stove was quite rusty.

The author's observations revealed that the biogas stove placed in his kitchen was indeed rusty due to his reluctance to clean it regularly after cooking. Moreover, he believed that biogas required no maintenance because, as he put it, "this [biogas stove] is given, so there's no need to care about it too much. If it breaks, it might be normal because it's not better than LPG" (Interview with DM). Although he considered asking the construction partner organization (CPO) to check and repair his biogas stove, he always canceled the request. Prior to the construction process, he had expressed great enthusiasm about the biogas installation in his backyard, believing that it would enhance his status as sub-village head. However, this enthusiasm was not matched by a commitment to care for the biogas system (Interview with DM).

Households that invest their own money in construction materials exhibit a strong sense of ownership and connection to their installed biogas systems. It is regrettable when users fail to maintain and care for their biogas units after making such significant financial investments, as expressed by the following housewife:

> To make our money worth, biogas needs daily careful maintenance, especially for the stove because it would be easily corrosive if not wiped after cooking. If the biogas stove broken or get rusty, I must spend a lot of money again and wait for couple months from the CPO. (Interview with SN)

It is noteworthy to confirm that household users who invest their own money in biogas installation tend to become more aware of their commitment to the technology. This investment signifies not only a financial commitment but also a dedication to the future sustainability of the biogas system. In other words, users' commitment to biogas may be strengthened by their awareness of the technology's function and benefits.

Utilizing leftover materials from home renovations—such as metals, bricks, and coral stones—has proven beneficial, leading to a cleaner environment and reduced construction expenses. One user, who previously overlooked the scattered metals from their post-earthquake renovation in 2006, came to recognize the advantages of reusing these remnants:

> While looking around my house, I noticed a lot of unused metal scattered about, remnants from the construction of another house after the earthquake. If I had purchased materials from a shop before installing the biogas system, it would have cost me a lot. Instead, I decided to reuse these materials, which saved me at least IDR 500,000. Now, I realize that sourcing our own materials not only provides significant financial benefits but also brings a sense of satisfaction. (Interview: RK 11.07.2011)

The geographical conditions present another challenge for users in maintaining the various components of biogas technology. One user is acutely aware that the biogas stove is susceptible to corrosion if not cleaned properly after cooking. This technical understanding has motivated her to pay close attention to maintenance, especially when it rains. SN, a housewife, acknowledged that humidity can adversely affect the performance of the biogas stove. Her concerns reflect a proactive approach to anticipating the impacts of changing weather conditions, which could diminish the value of the money invested in construction materials. Correspondingly, she recalled "Considering the costs our family has incurred, we feel obligated to clean and check the stove regularly, at least every three days. The humid climate in recent months has also prompted us to pay more attention to our biogas stove, as it might corrode due to the rain" (Interview with SN).

Previous experiences with broken pipelines have made users more proactive. Their frequent checks are driven by discussions with fellow users who faced unexpected issues, such as accidentally cut pipelines. Although these past experiences may have been traumatic, they serve as valuable lessons, enabling users to take preventive measures to avoid similar situations in the future. The money invested in construction materials also motivates users to care for their biogas systems through regular maintenance:

> After installation, and realizing how much money we spent, I prefer to check the biogas installation daily, and even more frequently. I heard that in other villages in Bantul, the gas couldn't be transported to the kitchen stove because local children stepped on the pipeline. I don't want to experience that. (Interview with RK)

From shared experience above, it is understandable that there has been a lot of subjective sense of ownership that user could develop by purchasing construction materials voluntarily as way of sustaining their biogas.

5.4.4 Supports of Family Members for Purchasing Construction Materials

Spending money on construction materials can be a significant consideration for household users, particularly for those with irregular monthly incomes. However, this situation does not diminish the potential for households to adopt biogas technology. Mobilizing resources within the family can serve as a strategy to overcome these financial challenges. The active engagement of family members in the financing process can diminish the costs of daily maintenance. When all family members share similar ideas and commitments, they can collaboratively find solutions. This is exemplified by the experience of NG's household. The family shared that a common approach among family members was key to addressing their financial issues. To finance the construction materials, which cost IDR 1,500,000 (approximately USD 150), the husband, wife, and son each contributed different amounts of money, thereby alleviating their financial constraints.

Their intention was also driven by the need to convert organic chicken waste, such as quills, into a more productive resource. This household's emphasis lies in their shared goal of integrating biogas technology while minimizing the ecological impact of chicken waste:

> My son contributed IDR 500,000, and my wife matched that amount as well. As the initiator in the family, I contributed the largest share of IDR 1,000,000. By doing this, we ensured that the responsibility for the biogas system is shared among all family members. Before we pooled our resources, we had a common concern about the excess chicken quills and how to convert them into a more productive asset for our household. (NG: Interview 07.07.2011)

The shared awareness and concerns about the chicken quills were also highlighted by NG's wife. She explained that the excess waste had motivated their family to seek alternative solutions that could add value to their business as chicken traders. Every day, they generate more than 10 kilos of quills. Prior to considering biogas, they simply covered the quills with soil for a week before disposing of them on their farm as organic compost.

5.4.5 Neighbor's Assistances for Initial Loading

After the biogas installation is completed, the user must load the reactor with an adequate volume of mixed dung and water to accelerate the initial gasification process. For a reactor with a size of 4 m^2, approximately 20 kg of dung is required for activation. Users often find it burdensome to gather this much dung solely from their own cattle, so they must explore other ways to meet this challenge and effectively activate their biogas plant. One user without cattle shared his experience with me about the difficult yet rewarding process of his first-time feeding. He transported kilos of dung using his pulling cart (*gerobak dorong*) from a collective cattle shed located 200 m away from his house, all under the heat of the midday sun and without any assistance from his children (Interview with NG). Despite the challenges, he found the experience exciting as it provided an opportunity for vibrant dialogue with fellow farmers about the benefits of biogas in keeping the cattle shed clean. A key takeaway from this story is that the long-standing trust within the community helped mitigate the difficulties he faced in collecting cattle dung for his household:

> This experience held significant meaning, as it allowed my fellow farmers to recognize the value of biogas, something they had only heard about on television. I believe this personal experience demonstrates that biogas can be effectively promoted through interpersonal encounters and daily interactions. As local farmers, we have established strong networks that have been maintained through community meetings, which is why I didn't need to ask for permission to collect dung from their cattle sheds. Trustworthiness among us is paramount. (Interview with TG)

In contrast, another user in the sub-village, who did not have access to collective cattle dung, shared a different experience. In his efforts to activate his biogas reactor, TG collected around 20–25 kg of dung by visiting the cattle sheds of five of his neighbors. Despite the exhausting effort involved in loading the dung into his carriage, it became evident that he had devised a strategy to persuade his neighbors to allow him to collect the dung. His success was largely attributed to the modest willingness and social trust he had established with those in his community.

> The night before I collected the dung, I visited five households with cattle to ask for permission first. This visit turned out to be invaluable, as it allowed me to brief them about my biogas reactor. All of them agreed to help, which made me very happy. I believe this kind of assistance is a form of social capital in our village and a way to sustain our social relations. (Interview: TG 09.07.2011)

If we examine the user's experience more closely, it becomes evident that he emphasized the importance of "sustaining social capital" with his neighbors. Providing assistance in this Javanese rural community is an integral part of social support, which has long been maintained as a means of fostering solidarity among its members (Sofyan et al., 2023; Syah, 2023).

5.4.6 Support of Kinship Throughout "Start-Up" Period

Biogas technology reinvigorates unique experiences for household users, particularly those with strong ties to siblings or family members. In my observations, one household user shared the biogas output by branching out the pipeline to another household. Technically, there are no restrictions on transporting gas output in this manner, as long as users are aware of the consequences, such as a reduced gas output.

In one case, a single biogas system was shared between two households. For the household with the installation, a 20-m pipeline was assembled to transport biogas to their kitchen, while another pipeline connected to the main reactor transported gas to the other house, which was only 10 m away. As a result, the main household's biogas volume was lower than that of the other household. Fortunately, since the second household was a family member, this arrangement did not create significant problems. Despite the unexpectedly lower biogas volume, it was found that maintaining a harmonious relationship and avoiding conflict was an effective solution to the issue. ME, a housewife and biogas user, willingly accepted the smaller gas volume to sustain her connection with her elder sister:

> Since the first installation, I agreed to branch the pipeline from the reactor because I wanted to share the gas with my brother. However, the situation is not as smooth as I had planned; the biogas flow to my sister's stove is much better than to mine. Initially, I thought that our $6m^3$ volume would provide an excess of gas. But now, there's nothing I can do about it. As the younger sister, I view this as a positive way to enhance my emotional ties with my brother. I don't regret my decision at all. (Interview with ME)

In addition, her statement implied that she felt no regret about her unanticipated plan; instead, she viewed this experience as an opportunity to strengthen family ties with her sibling, who lived just five meters away. Fortunately, her husband supported her decision. He emphasized that preserving family values in Javanese life takes precedence over any technical decision because, after their parents pass away, siblings are the ones who can be relied upon for assistance (Interview with BW). This household's experience highlights, that family values play a significant role in altering the functionality of biogas technology, even when the outcomes may be unpredictable. This finding is essential for the development of biogas systems, as solidarity and the desire to maintain family ties are woven into the use of the technology itself, making such relationships indispensable.

The author also discovered that kinship relationships serve as invaluable sources of information among households. On different occasions, kinship ties may circulate biogas as a social symbol of privilege within Javanese families. The Javanese way of life is deeply symbolic, often not easily understood by outsiders. Although everyone is treated equally, showcasing family wealth can be perceived as ostentatious. In line with this perspective, a laborer working with biogas remarked:

> The family connection between MI and the wife of SS is one of the facilitating factors that increases the likelihood of biogas adoption for both families. Among her neighbors, MI is seen as one of the wealthiest women because her husband is a cattle broker who earns a substantial income. Thus, sharing ideas with neighbors of differing wealth levels can be somewhat reluctant. (KM: Interview 29.07.2011)

For first-time adopters, using a biogas stove can be more labor-intensive, as it requires a thorough understanding of its technical complexities. This contrasts with the more user-friendly LPG stove, which most users are already accustomed to (Berruto et al., 2013; Lessen & Sorensen, 2006; Parawira, 2009; Wang et al., 2016). Operating the biogas stove necessitates pushing a switch in the right direction to turn it on; only then can cooking commence. Many housewives find the manual instructions confusing, especially if they are used to the LPG stove. This was also the case for the wife of SS during the first month following the completion of her biogas installation. BW mentioned that she often called SS's wife for advice on time management and cooking preparations with the biogas system.

> Sometimes, I share with her that we must wake up earlier to manage the biogas and check the manometer to monitor the gas volume. If there's no gas, our cooking activities will be interrupted, which can ruin our day! If the water indicator on the manometer rises, it indicates that there's still gas available

5.5 Objectifying Biogas

When biogas users have the ability to use and incorporate habits into their routines, they will do their part to keep the biogas technology compatible with their habits, work, living situations, farming activities, and so on. This section details all of these social practices from the everyday narratives of the biogas users.

5.5.1 Employing Labor for Feeding Routines

Feeding the biogas reactor with an adequate mixture of dung and water is crucial for its operation and is primarily dependent on the user's role as an actor. While the slurry, consisting of a 1:1 ratio of dung to water, is loaded every three days based on the reactor's size, any failure to maintain this schedule can lead to interruptions in biogas production and degrade its performance. To ensure their biogas systems remain operational, some users take proactive steps to maintain sufficient slurry levels, often negotiating these tasks with their daily routines. Employing paid labor is one option for those who can afford it, indicating that biogas technology is more readily accepted among middle-income farmers (Alam, 2013). For instance, SS, a busy policeman, has arranged for a nearby neighbor to assist him with loading the mixture every three days due to his demanding work schedule.

> Although I am quite busy at work and back home around 5 pm every day, I always remember to take good care of biogas. The possible way I can do is asking my neighbor who always weeds every noon at my yard to feed the reactor with dung taken from my cattle shed. I am so lucky because he never asked me for money. (Interview with SS)

Unlike SS, another user, SG, who is frequently busy with work, has to spend extra money to hire labor for maintaining the biogas system. SG explained that he employs paid labor because he expects them to be dedicated to the regular feeding process, having received prior training for this task (Interview with SG, 24). As the local village head and a well-known figure due to his successful furniture business throughout the district, SG's financial capability to hire labor is an enabling factor. Furthermore, he has a strong desire to be viewed as a role model for other household users committed to the sustainability of biogas.

BW's experience is somewhat similar to SG's. By paying IDR 50,000 per week for labor, he ensures a steady flow of biogas to his kitchen (BW: Interview 30.07.2011). However, his wife, who runs an alternative medical practice from home, is reluctant to operate the biogas reactor, viewing it as a task suited for men.

5.5.2 Negotiating with New Habit and Tackling New Smell of Dung

The shift from using LPG to a biogas stove resulted in a distinctive experience for users. Informants I interviewed shared similar insights regarding this transition period. This phase often involved specific coping strategies and efforts to integrate the new technology into their daily routines. The process during this time can be characterized as experimental learning; users attempted to tame the new technology by engaging in various experiences and rationalizing them with scientific explanations that offered logical answers. What initially appeared strange, unpleasant, or unfamiliar eventually prompted a rational understanding of the technological artifact. To illustrate how users learned to adapt to the unfamiliar tastes associated with biogas cooking, one informant recalled:

> In the first week of installation, I smelled a very strong odor of cow dung that lingered for quite a while. The smell made me nearly vomit. After several days of use, I realized that it was the sulfur and methane from the biogas being released for the first time. Thankfully, I managed to tolerate that smell for up to seven days. (Interview with NT)

Through this account, NT expressed her genuine experience with the biogas stove, which she had never used before. The introduction of the biogas stove posed a significant challenge in her role as a housewife. She interpreted the unpleasant odor emanating from the stove not as a result of cow dung, but rather as methane that had accumulated in the biodigester and was being released for the first time. This rational understanding emerged from her prior experience of cleaning the cowshed, which produced a similar smell.

The construct of "smell" established by users can vary widely. Self-assurance and confidence are strategies they employed to navigate the differences in taste, helping them through this transition stage. During a visit to a local household, the head of the household offered me two glasses of water and insisted that I drink both. Afterward, he asked for my assessment of the taste. I replied that both tasted the same. He then revealed that one of the glasses contained water boiled on the biogas stove. He explained:

> …the water boiled on the biogas stove tastes just the same and very good. So, I believe that the bad smell of cow dung is merely our perception of the flavor when cooking with biogas. Essentially, what we smell is influenced by our imagination of what we are consuming. (Interview with RK)

His explanation highlights that individual perceptions of food can significantly affect the perceived taste. This understanding requires users to adapt to unexpected experiences, converting them into new, harmonious interactions with the technology.

5.5.3 *Innovating the Cowshed*

Innovation among users enhances the convenience of biogas utilization and simplifies their tasks related to biogas management. One notable innovation is the integration of the inlet with the cowshed. Unlike conventional inlets, the integrated system allows users to shovel dung quickly and mix it without needing to pick it up manually, saving time and increasing convenience. During the domestication of this technology, innovation reflects users' expectations and manifests their desire for biogas technology to harmonize with their existing customs and habits. For those who care for cows, they can feed the biogas reactor with a blend of water and dung while simultaneously cleaning the shed, eliminating the need for redundant tasks.

Interestingly, users have the freedom to modify their installations according to their needs. Although the construction partner organization (CPO) recommended placing the mixer and biogas plant at the back of the home near the kitchen, users are not restricted from implementing their own ideas. Many users integrate their routine cleaning of the cowshed, which they typically do each morning, with feeding the biogas reactor to save time and enhance convenience. Below are illustrations (see Figs. 5.4, 5.5, and 5.6) of the integrated system connecting the cowshed and the biogas mixer. As shown, users created a channel through a pipeline from the cowshed to the mixer. After cleaning the shed, they no longer need to transfer dung into a bucket a process that is time-consuming and inefficient. Instead, they simply spray it with water and shovel it directly to the mixer.

Fig. 5.4 The users showing me his integrated inlet. The model of the mixer is kept as it is but he made it with lower surface to adapt with upper position of the cowshed (*Source* Author)

Fig. 5.5 A user shoving the cow dung into the inlet and then adding it with water to make it more aqueous (*Source* Author)

Fig. 5.6 The integration of the biogas inlet with the cowshed eliminates delays in feeding the reactor with water, enhancing comfort and convenience for users (*Source* Author)

During observations, the author noted three households employing this integrated cowshed system. The stories behind these innovations are compelling. For instance, every morning, SD juggled his preparations for work while also needing to clean the cowshed to maintain the health of his cows. This modification benefited not only him but also his wife, who worked as a trader at a nearby traditional market. When I visited their home, his wife expressed her excitement over the innovation. After SD left for the office, she efficiently fed the biogas plant with the mixture. The innovation he implemented simplified her tasks, allowing her to save time by merely rotating the mixer's handle and channeling the mixture through the pipeline to the biogas plant.

Another user, SL, who lives in a different sub-village, exhibited a similar pattern. In this household, the mixer and biogas plant are located just seven meters from the dining room and kitchen. This close proximity was intentional, allowing for easy access when checking the available volume of gas. In the following (see Fig. 5.4) mixture of cow dung and water go through to the lower channel of the mixer. By such way he can keep the hygiene of himself and clean the cowshed.

This innovation highlights the importance of daily household routines, as the placement of the mixer near the cowshed significantly affects efficiency. In contrast, other households, aside from those of SL and SD, follow a standard design where the cowshed and mixer are separated by 3–5 m. This distance requires users to expend more time and energy picking up dung, ultimately making their work more labor-intensive.

The adoption of this new household technology inevitably leads to the emergence of new habits among users. However, most users, particularly the male household members, do not feel burdened by the need to schedule regular feedings for the biogas plant or digester every few days. Their familiarity with agricultural tasks involving dung and animals means that this new habit integrates seamlessly into their daily activities without hindering other responsibilities. Moreover, the role of technology in stimulating innovation is immense. As highlighted in Tornel's (2022) study, this technical self-empowerment decolonizes the innovation process, allowing locally generated ideas from biogas users to emerge and ensuring that solutions are tailored to meet local needs and conditions. Additionally, Dunlap's (2020) critical assertion emphasizes that grassroots innovation aligns with the sustainability ideals of the technology itself. Users are increasingly recognizing the importance of practices that promote social equity and environmental stewardship. By implementing authentic initiatives, these methods enhance the community's resilience to the potential limitations of biogas technology.

In summary, the significance of innovation for local communities extends beyond economic metrics. It encompasses social, cultural, and environmental dimensions that collectively enhance quality of life and community resilience (Cruz, 2021). By embracing innovation, local communities are fostering a sustainable, inclusive, and vibrant future, thereby contributing to a more equitable society (Meighan, 2021; Murphy & Largacha-Martínez, 2022; Wagner & Fernández-Ardèvol, 2019).

5.5.4 The Interplay of Gender Relations in the Use of Biogas

Before biogas can be effectively utilized in households, several stages, including construction, must be completed. Although organizations like Biogas Biru and AWFDFFM provide construction labor and subsidized galvanized pipelines to connect the biogas reactor to the stove, each biogas user is responsible for purchasing construction materials at their own expense. This approach not only reduces overall costs but also allows users to save significant amounts of money. In determining their financial sources for these construction materials, users exercise autonomy over their spending. Once the biogas design is finalized and the necessary materials are acquired, the construction process can begin. Throughout this process, gender roles and the division of labor play a crucial role within the socio-technical system or network that influences the domestication of technology. Consequently, understanding the comparative involvement of women and men in this context becomes particularly relevant (Bittman et al., 2004). Figure 5.7 illustrates the involvement of women and men in biogas construction. Interestingly, we observe a paradoxical situation: while biogas is appropriated, men emerge as the dominant actors throughout the construction process, from initial contact with the construction partner organization (CPO) to the mobilization of construction material.

In Javanese rural families, after marriage, women traditionally move into their husband's house as a demonstration of loyalty. The land property is typically owned by men, which can lead to women's lack of robust standing in decision-making processes related to financial sources, signing agreements, and selecting the reactor size for biogas technology. In interviews with female informants, many revealed that while their husbands offered them a chance to negotiate aspects such as the reactor size and financial resources for purchasing materials, they often felt powerless to voice their opinions. This powerlessness stems from their husbands having already mastered the necessary knowledge about construction techniques. This observation raises an interesting point regarding how technological artifacts are associated with masculine identity. As Omrod (2006: 47) noted, "sometimes it is seen as the maintenance of masculine dominance around notions of greater sophistication and knowledge content of the technical in contrast to the family as a feminized domain with its everyday life."

However, women in these households do demonstrate authority over certain aspects of domestic life, such as the arrangement of the kitchen, food preparation, and child-rearing. When it comes to the placement of technological artifacts, women assert well-defined decisions that impact the future sustainability of the housing infrastructure. Female participants in this study expressed that although they were not actively engaged in the construction process of biogas technology, their involvement was linked to their moral responsibility for maintaining the home as a space for domestic activities. For instance, one participant, SF, articulated this perspective:

> I decided that the biogas stove should be placed outdoors, not in the indoor kitchen. It would damage the painted wall and require perforating it, which would make our home look less aesthetic.

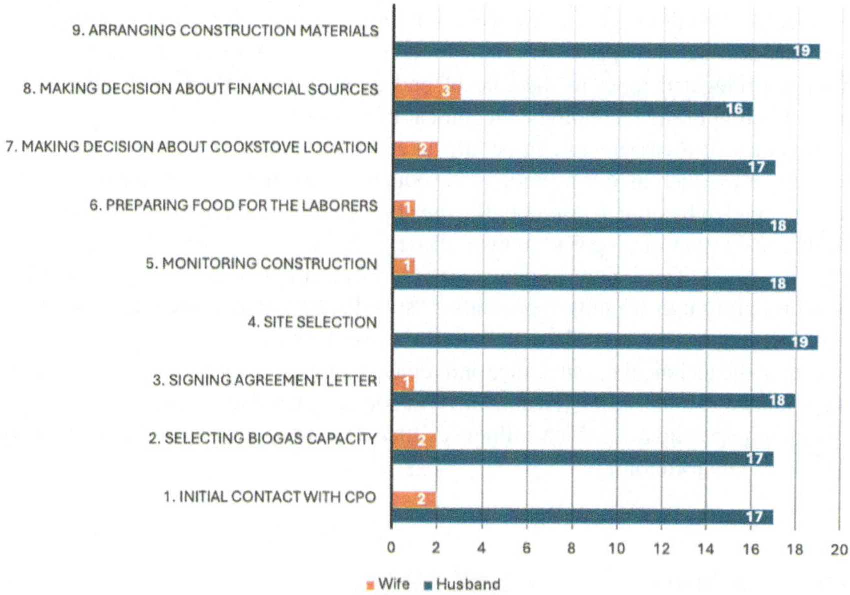

Fig. 5.7 Bar chart showing the division of responsibilities between a husband and wife in various construction-related activities. The chart lists nine tasks such as arranging construction materials, making decisions about financial sources, and preparing food for laborers. Each task bar is split to show the number of tasks undertaken by the wife (orange) and the husband (blue), with the husband taking a significantly larger share in most activities except in making decisions about financial sources and selecting biogas capacity, where the wife's involvement is slightly more (*Source* Author)

SF's statement reflects a wife's concern for the aesthetic arrangement of their home environment. This perspective aligns with Omrod's (2006) assertion that women possess knowledge related to taste and bear the responsibility for home decoration. Furthermore, it highlights the expectation that women, upon moving to their husband's residence, should preserve the man's property.

The disengagement of women from the biogas technology process is intriguing, as it reveals the intertwining of masculinity and femininity within technological artifacts (Akrich, 1992). Biogas technology is often perceived as technical and linked to physical strength, leading women to believe that acquiring technical knowledge is not within their rights. One wife, ME (a pseudonym), expressed this sentiment when discussing her views on biogas technology:

> Despite receiving no sufficient knowledge about biogas technology compared to my husband, I think that is okay for me because technique— related issues are under the responsibility of a man, not woman. I decide the kitchen thing, and he take care of technical thing like electricity and the likes.

Nevertheless, women cannot be blamed for their self-separation from the technical aspects of biogas technology. Many interviewees noted that wives primarily received technical information and knowledge about biogas from their husbands, rather than from the construction partner organization or local training programs. This situation has relegated them to the role of end-users, leaving them with insufficient understanding of the technology's technical functions.

This lack of involvement can be attributed to the limited presence of women in public spaces, such as during biogas introductory meetings, which were predominantly attended by men. Consequently, the absence of women in the technical maintenance and repair of biogas systems is closely tied to what is termed "social script" (Akrich, 2002). The failure to engage women from the outset of biogas socialization through to user training represents a "social script" that shapes their exclusion from the socio-technical tasks associated with this technology. Thus, the absence of women in the technical maintenance and repair of biogas systems is directly linked to the product's "social script" (Hesni, 2024) or social values that instructs an individual to undergo specific act, which influences their lack of inclusion in the operational aspects of this technology.

5.6 Conclusion

In making biogas technology acceptable and integrated into household spaces and environments, two levels of influence are at play: the macro-situation, which involves dynamic social interactions between potential users and technological artifacts facilitated by responsible actors such as construction partner organizations, and the micro-situation, which focuses on specific household contexts where users must adapt their eating habits, try new tastes, mobilize relentless efforts with neighbors, and maintain an open-minded attitude toward adopting this technology. Compared to other common cooking stoves that are simpler and more instantaneous, biogas technology requires users to acquire substantial knowledge regarding technical maintenance and a dedication to learning the intricacies needed for seamless integration into their daily lives. The findings suggest that the introduction of biogas technology characterizes a dialectical process, meaning that the technology must be compatible with local settings and cultures, while users must cultivate enabling social systems that support the sustainability of the technology's use.

Overall, these findings confirm the assertion of Mackay and Gillespie (1992) that technological artifacts become functional when they facilitate socio-technical objectives that integrate interactions among non-government actors, family members, and even non-users. In line with this assertion, Hynes and Richardson (2009) argue that both users and non-users engaged in the process of objectification play roles as actual and potential consumers seeking the desired object. This study reveals that users' preferences for biogas technology are shaped by an interplay of social networking, introductory meetings, and online media, while internal solidarity and commitment within households help them overcome financial constraints when purchasing construction materials. Additionally, kinship bonds play a prominent role in exchanging alternatives for coping with technical issues. Given that biogas is used domestically within households, gender roles are deeply intertwined with its adoption. The findings indicate that men still dominate public participation in accessing technological knowledge and in decision-making processes regarding the acquisition of biogas technology.

Before moving further, it is essential to understand that the acquisition of biogas technology in this study is shaped by the knowledge and information, as well as the social networking among users and non-users. The channels through which users access this technology profoundly influence their preferences and capabilities in constructing biogas systems. In the context of objectifying biogas technology, Subedi et al. (2014) highlight that technological awareness, informed by information from various channels, is crucial for altering and molding user preferences.

Before adopting biogas technology, potential users visualize how it operates and how it is technically maintained, but this knowledge is often insufficient. Thus, demonstration plots are necessary to bridge this gap. The findings indicate that biogas users have been exposed to demonstration plots as introductory sites, either through direct visits or documents. Such experiences facilitate firsthand engagement with the technology, which is significant for generating initial familiarity. In relation to the importance of demonstration plots and their influence on user preferences, domestication experts argue that new introductory practices and organized events for improving technological trialability can help users mentally construct their understanding of the technology (Floyd et al., 2003). This kind of cognitive processes involved in learning new practices are vital for acquiring new meanings that lead to the acceptance of technology. Similarly, Floyd et al. (2003) assert that awareness and time allocated for experimenting with the technology can significantly impact decision-making. The users' previous experiences with similar technologies also play a crucial role in shaping their preferences, as Haddon (2006) emphasizes that a history of engaging with former types of biogas profoundly influences user inclination toward biogas technology.

Transitioning to a newly adopted technology, such as moving from an LPG-based cookstove to a biogas cookstove, involves negotiations with pre-existing values, tastes, and routines that users have maintained. Reidsma et al. (2006) affirm that adjusting to new tastes, values, and styles through the use of new technology is essential for a smooth transition and to avoid conflicts. Furthermore, taste is linked to home decoration and aesthetics, which are often more gendered than other aspects of technology adoption. Responding to changing habits, Akrich (2002) argues that technology can generate and naturalize new forms of causality and new knowledge about the world through negotiated tastes and the performance of new habits.

This chapter has elucidated that acquiring biogas technology involves various dimensions that facilitate this process, including not only knowledge and financial aspects but also social networking with NGOs. Despite the lack of a clear boundary between appropriation and objectification, the discussion highlights that the needs, desires, and intentions of individuals and groups are dynamically attributed dimensions that inevitably shape the use of biogas technology.

References

Akrich, M. (1992). The description of technical objects. In W. E. Bijker & J. Law (Eds.), *Shaping technology/building society: Studies in sociotechnical change* (pp. 205–224). MIT Press.

Akrich, M. (2002). An experiment in techno-sociology. In P. Lemonnier (Ed.), *Technological choices: Transformation in material cultures since the Neolithic* (pp. 289–299). Routledge.

Alam, M. (2013). Theorizing the social acceptance of appropriate technological. *Deleted Journal, 11*(1), 81–92. https://doi.org/10.14203/stipm.2013.13

Berruto, R., Busato, P., Bochtis, D., & Sørensen, C. G. (2013). Comparison of distribution systems for biogas plant residual. *Biomass and Bioenergy, 52*, 139–150. https://doi.org/10.1016/j.biombioe.2013.02.030

Bittman, M., Rice, J. M., & Wajcman, J. (2004). Appliances and their impact: The ownership of domestic technology and time spent on household work. *The British Journal of Sociology, 55*(3), 401–423. https://doi.org/10.1111/j.1468-4446.2004.00026.x

Cruz, C. C. (2021). Decolonizing philosophy of technology: Learning from bottom-up and top-down approaches to decolonial technical design. *Philosophy & Technology, 34*, 1847–1881. https://doi.org/10.1007/s13347-021-00489-w

Dunlap, A. (2020). The direction of ecological insurrections: Political ecology comes to daggers with Fukuoka. *Journal of Political Ecology, 27*(1). https://doi.org/10.2458/v27i1.23751

Floyd, C., Harding, A.-H., Paudel, K. C., Rasali, D. P., Subedi, K., & Subedi, P. P. (2003). Household adoption and the associated impact of multiple agricultural technologies in the western hills of Nepal. *Agricultural Systems, 76*(2), 715–738. https://doi.org/10.1016/s0308-521x(02)00152-x

Haddon, L. (2006). The contribution of domestication research to in-home computing and media consumption. *The Information Society, 22*(4), 195–203. https://doi.org/10.1080/01972240600791325

Hesni, S. (2024). How to disrupt a social script. *Journal of the American Philosophical Association, 10*(1), 24–45.

Hynes, D., & Richardson, H. (2009). What use is domestication theory to information systems research? In *Handbook of research on contemporary theoretical models in information systems* (pp. 482–494). IGI Global.

Joshi, R. G., Chelliah, J., & Ramanathan, V. (2015). Exploring grassroots innovation phenomenon through the lived experience of an Indian grassroots innovator. *South Asian Journal of Global Business Research, 4*(1), 27–44.

Lessen, E., & Sorensen, C. (2006). Integrating: Technology in schools, colleges, and departments of education: A primer for deans. *Change: The Magazine of Higher Learning, 38*(2), 44–49. https://doi.org/10.3200/chng.38.2.44-49

Lewanika, M. M., & Mulenga, K. D. (1996). Constraints on the development of biotechnology in Zambia. *World Journal of Microbiology and Biotechnology, 12*(5), 463–465. https://doi.org/10.1007/bf00419458

Mackay, H., & Gillespie, G. (1992). Extending the social shaping of technology approach: Ideology and appropriation. *Social Studies of Science, 22*(4), 685–716. https://doi.org/10.1177/030631292022004006

Mazunda, J., & Shively, G. (2015). Measuring the forest and income impacts of forest user group participation under Malawi's forest co-management program. *Ecological Economics, 119*, 262–273. https://doi.org/10.1016/j.ecolecon.2015.09.016

Meighan, P. J. (2021). Decolonizing the digital landscape: The role of technology in Indigenous language revitalization. *AlterNative: An International Journal of Indigenous Peoples, 17*(3), 397–405. https://doi.org/10.1177/11771801211037672

Mshandete, A. M., & Parawira, W. (2009). Biogas technology research in selected sub-Saharan African countries—A review. *African Journal of Biotechnology, 8*(2). https://doi.org/10.4314/ajb.v8i2.59749

Murphy, J. W., & Largacha-Martínez, C. (2022). Decolonization of AI: A crucial blind spot. *Philosophy & Technology, 35*(4). https://doi.org/10.1007/s13347-022-00588-2

Omrod, S. (2006). 'Lets nuke the dinner': Discursive practices of gender in the creation of a new cooking process. In R. F. Dilic (Ed.), *Bringing technology home: Gender and technology in a changing Europe.* Open University Press.

Pal, J. (2003). The developmental promise of information and communications technology in India. *Contemporary South Asia, 12*(1), 103–119. https://doi.org/10.1080/0958493032000123399

Parawira, W. (2009). Biogas technology in sub-Saharan Africa: Status, prospects and constraints. *Reviews in Environmental Science and Bio/technology, 8*(2), 187–200. https://doi.org/10.1007/s11157-009-9148-0

Reidsma, P., Tekelenburg, T., van den Berg, M., & Alkemade, R. (2006). Impacts of land-use change on biodiversity: An assessment of agricultural biodiversity in the European Union. *Agriculture, Ecosystems & Environment, 114*(1), 86–102. https://doi.org/10.1016/j.agee.2005.11.026

Sofyan, M. A., Wahyuni, T., & Arifin, W. L. (2023). Javanese Islam and globalization: A study on the sustainability the Aboge Islam community in Indonesia. *Jurnal Sosiologi Reflektif, 17*(2), 369–400. https://doi.org/10.14421/jsr.v17i2.2690

Subedi, M., Matthews, R. B., Pogson, M., Abegaz, A., Balana, B. B., Oyesiku-Blakemore, J., & Smith, J. (2014). Can biogas digesters help to reduce deforestation in Africa? *Biomass and Bioenergy, 70*, 87–98. https://doi.org/10.1016/j.biombioe.2014.02.029

Syah, I. (2023). Social solidarity of Javanese society in seven villages of Modayag district. *Kawanua International Journal of Multicultural Studies, 4*(2), 131–138.

Tornel, C. (2022). Decolonizing energy justice from the ground up: Political ecology, ontology, and energy landscapes. *Progress in Human Geography, 47*(1), 43–65. https://doi.org/10.1177/03091325221132561

Wagner, S., & Fernández-Ardèvol, M. (2019). Decolonizing mobile media: Mobile Internet appropriation in a Guaraní community. *Mobile Media & Communication, 8*(1), 83–103. https://doi.org/10.1177/2050157918822163

Wang, C., Zhang, Y., Zhang, L., & Pang, M. (2016). Alternative policies to subsidize rural household biogas digesters. *Energy Policy, 93*, 187–195. https://doi.org/10.1016/j.enpol.2016.03.007

Chapter 6
Incorporating and Conversing Biogas

Abstract The integration of biogas technology into the daily lives of families, particularly within their established routines, is essential for its success. This chapter presents an analysis of how households adapt their patterns to fully incorporate biogas technology into their everyday activities. It also examines the actual experiences of families with this integration and its subsequent impact on their lives. Biogas technology can seamlessly fit into family life when it aligns with the established routines and long-standing habits that family members have cultivated. Changes in the spatial organization of homes and the reallocation of time reflect how users' perceptions of past realities interact with the introduction of this new household technology. Moreover, the integration process is closely linked to pre-existing consumption behaviors that are culturally ingrained within the home. These findings support the assertion made by consumption scholars that renewable energy practices can extend beyond the religious and cultural frameworks of a society, influencing its utilization on a broader social level. In this chapter, the domestication framework of integration and conversion elucidates how technology becomes extensively woven into the fabric of society.

Keywords Biogas · Incorporating · Users · Homes · Renewable energy · Domestication

6.1 Incorporating Biogas into Everyday Life

Biogas must be incorporated into family life, particularly within their established routines. This chapter discusses the changes in household practices that facilitate the seamless integration of biogas technology. It also examines how families experience this integration and the subsequent transformations in their daily lives. Biogas technology can be fully assimilated into family life when it is harmonized and synchronized with the routines and long-standing habits that family members have maintained. The alterations in the spatial organization of homes and the arrangement

M. Alam, *The Social Life of Biogas*, https://doi.org/10.1007/978-981-96-3300-5_6

of time indicate that users' perceptions of previous realities in relation to this new household technology have taken shape (Schmid & Smith, 2020).

6.1.1 The Arrival of Food Preparation and Time Arrangement in the Family Life

The introduction of biogas technology into households has led users to share and explore new concepts of time and temporality. Time is a hallmark of modernity, shaping the reasoning of contemporary society by limiting, framing, and regulating individual activities (Fritzsche, 2001). Each activity's productivity is marked by material boundaries defined by hours and minutes (Swedberg, 2011). Before the availability of biogas, households did not concern themselves with time management; they could cook indefinitely using an LPG burner. However, after implementing biogas, users became more conscious of their time usage.

The duration spent on cooking varies significantly among households, depending on their daily cooking routines: mornings (4:00–9:00 a.m.), afternoons (11:00 a.m.–3:00 p.m.), and evenings (5:00–7:00 p.m.). Initially, households assumed that the biogas volume would last for all-day cooking. However, with normal feeding intervals of once every three days, the biogas typically lasts between five days to a week, provided that users cook for a maximum of six to seven hours daily. Interviews revealed that households recognize the need to manage their cooking times, as excessive use—beyond six to seven hours a day can quickly deplete the gas supply. This normalization illustrates that users understand the renewable nature of biogas while acknowledging its temporal limitations in usage and capacity. Consequently, biogas users must restrict their cooking time to conserve gas. One strategy is to use a gasoline burner alongside the methane cookstove to ensure that the biogas supply does not diminish too rapidly. This social practice highlights the coexistence of traditional and modern energy sources in daily use.

Pierre Bourdieu's (1963) research in Algeria illustrates a similar transformation. Before the advent of efficient agricultural technologies, Algerian farmers relied on circadian rhythms to predict and ensure harvest periods. This reliance evokes a sense of anticipation rooted in past experiences, as Giddens (1991) noted, without colonizing the future with plans or ambitions. However, the introduction of new technology requires farmers to become more anticipatory with their time throughout the day, prompting them to engage in conscious planning (Bourdieu, 1986). Most households also used their biogas stoves to prepare meals for farm laborers who assisted with harvesting paddy, finding biogas sufficient as the main fuel source. Compared to LPG stoves, biogas stoves require less energy, but users recognized that operating the biogas system was different. The biogas stove depends on maintaining a consistent methane gas supply, which requires regular feeding with cow dung. However, since the operation of a biogas stove—such as rotating the switch to control the flame—closely resembles that of an LPG stove, users reported no significant difficulty in adapting to it. Their familiarity with LPG stoves made transitioning to biogas stoves relatively easy. This experience demonstrates that their use of biogas stoves was informed by their previous knowledge of cooking technology. As most users had long relied on LPG stoves, they encountered little difference when using biogas stoves. This aligns with the concept that old technologies are not abandoned but integrated into the optimization of new ones for household use (Rizzi et al., 2014). The familiarity gained from previous technology supports the adoption and use of renewable energy systems like biogas (Popp et al., 2011).

Introducing new technology into the household can challenge existing habits. These habits, formed over time through the use of older technology, are negotiated, contested, and repeated, but not entirely discarded. As Shove (2012) notes, both old and new routines coexist, being "recurrently and consistently reproduced by suitably committed practitioners" (Shove, 2012: 103). In this case, the introduction of the biogas stove did not eliminate the use of other cooking methods. Many users continued to use their LPG stoves, but for different purposes. For example, households that own cows must feed their livestock every morning with about five liters of rice chaff before the husband leaves for work, while the wife prepares breakfast on the biogas stove. In some cases, they still used the LPG stove to speed up cooking the rice chaff, as it was considered faster than using the biogas stove for that task.

> If I use biogas for my cow, I will be late to reach the office because I must wait for the rice chaff to cook, which takes more than 30 minutes. So, we prioritize biogas only for making breakfast, as our breakfast menu is simpler; we usually just simmer the dishes we prepared the night before. (Interview with SD)

This statement highlights users' rationality in evaluating the effectiveness of biogas technology, viewing it not as a taken-for-granted renewable energy source. In today's modern era, as illustrated by this experience, the notions of effectiveness and efficiency as time-bound cultural practices are "historically and socioculturally specific phenomena, arising from distinct sociocultural engagements with locally particular processes of modernity" (Musharbash, 2007: 37).

Consumption patterns vary from one household to another, shaped by the specific farming activities they engage in. For households employing more than one farm laborer, biogas is often considered insufficient for cooking. For example, in a household with three laborers working in their paddy fields from morning to afternoon, extra stoves are necessary, leading them to choose LPG stoves. Although there are only two children in the family, the household head believes cooking on biogas alone is inadequate for preparing breakfast, as they need to provide extra food for the laborers. He noted,

> My two growing children, aged 3 and 12, only need very simple meals, but the construction workers do not. We have to serve them a variety of dishes since they work until the afternoon, while my wife is at work and cannot cook for them (Interview with TG).

This experience illustrates that external consumption, which does not directly relate to the family's domestic needs, also influences household energy use. Such patterns are linked to the agricultural activities they undertake during a given month or period. This case underscores the notion that biogas, as a renewable energy source, does not always meet the needs and desires of consumers (Wallenborn & Wilhite, 2014).

6.1.2 Time Adjustment and Reorganizing Space in Family

The transition to biogas stoves is relatively smooth; however, it does not guarantee that users can shift to using biogas without leaving behind their previous cooking methods, such as using LPG. In this context, Winther (2008) argues that the habit of using former cooking equipment may lead users to experience a transitional period during which they engage in experimental practices to understand the distinct realities of using new technology. For new adopters of biogas, changing habits becomes an embedded aspect of their daily routines. This is particularly true for households transitioning from LPG gas stoves to biogas stoves. Before adopting biogas, many users—especially women who frequently engage in cooking—found LPG stoves much easier to operate. With LPG, they simply turn the "ON" button, and the blue flame ignites instantly, making the cooking process quick and efficient. In contrast, biogas stoves are designed differently; users must first light the burner cup and then wait a few minutes for the flame to become stable before they can start cooking. This design prioritizes user safety, requiring users to check the manometer to gauge the remaining gas supply and allocate time for cooking accordingly. Interviews revealed that users generally need to allocate extra time—usually between 2 and 5 min—before they can begin cooking.

In response to that, the presence of gender hierarchy in technological innovation is significant. Following Gonda (2016) and Uchenna and Oluwabunmi (2020), cooking technology has long been associated with domestic roles for women in the Global South. In daily household use, housewives are required to check the manometer frequently before preparing meals. Initially, they may feel hesitant and reluctant to perform this task, especially during the first month of using biogas technology. However, by the second month, they tend to acclimate to the new routine.

With the emergence of biogas technology, there is a heightened demand for household users—especially housewives—to understand essential technical knowledge related to daily use, such as how to determine whether the gas volume is sufficient and what indications on the manometer signify the availability of gas. Most users acquire this technical knowledge through direct instruction from their husbands, who often possess a better understanding of the system. This knowledge gap exists because husbands typically gather firsthand information from the technicians responsible for the post-construction phase of biogas development. Despite this knowledge disparity, the willingness of women to learn after adopting biogas indicates their desire to integrate it into their daily cooking routines for their families. a woman respondent asserted.

> How would we know how to care for it if we don't investigate the concepts of manometers, filling the intake with cowdung and water, and the piping system for distribution from the biogas digester to the kitchen?. The spouse cannot be hanged since he is not always at home.

This remark highlights that home biogas technology empowers women to acquire what Pakulski refers to as "cultural citizenship" (1997: 73), positioning themselves equally with men in the technical care and maintenance of these systems. In this context, cultural citizenship signifies the tangible efforts, initiatives, and resilience demonstrated by women in asserting their symbolic presence as "managers" of the mechanics associated with domestic technology. More broadly, this phenomenon suggests that home technology has transformed the gender dynamics within users' domestic spaces (Fraune, 2015; Risman & Davis, 2013).

Traditionally, women in rural areas have been granted authority over home care and the decoration of living spaces, despite land ownership typically residing with their husbands. Compared to men, women often exhibit greater concern for the interior aesthetics of their homes (Bensch & Peters, 2015; Kumar & Mehta, 2016; Malhotra et al., 2004). While this trend cannot be generalized, it is significant to explore how women often show a deeper attention to domestic spaces. For instance, SD's wife expressed dissatisfaction with her husband's decision to install the biogas stove in the kitchen, as illustrated in the following account:

> Well, my husband told me he wanted to drill the wall to install the pipeline for the gas stove, but I believe that would compromise the aesthetics of our home. I think no woman would want the walls of her house to be perforated; it would diminish our home's beauty. It was a big no for me. (Interview with SS)

On the other hand, families often designate specific areas in their homes for gathering and socializing, which are considered essential living spaces. The reorganization of these spaces following the adoption of biogas technology signifies a shift in meaning. This reflects organizing practices that serve as transformational outputs and negotiations of the socio-cultural boundaries present in individuals' experiences. This phenomenon aligns with the observations made by anthropologists Alberto Arce and Norman Long:

> a reworking of modernity values and practices takes place through the ways in which various social actors and groups process and act upon their experiences, thus re-constituting or transforming existing 'localised' situations, cultural boundaries and knowledge. This often results in the opposition or negation (not always directly) of the culture and knowledge of the expert or intervenor in favour of well-tried local ways and understandings, and through the appropriation and reinterpretation of modern idioms, technologies and organising practices. (Arce & Long, 2003: 7)

For further illustration, the author present observations from the household of WJ. Figure 6.1 depicts the room organization before adopting biogas, while Fig. 6.2 shows the reorganized space, indicated by the relocated living room closer to the kitchen.

As illustrated in Fig. 6.2, the family traditionally gathered in the living room located in the large corridor outside, which served as both a dining area and a space for casual conversations during warm evenings spent together. The glass windows built into bedrooms 2 and 3 allowed family members to keep watch over their cows.

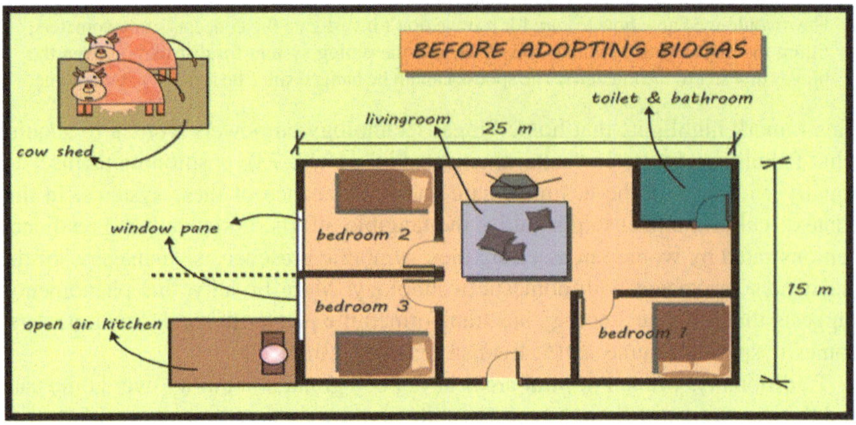

Fig. 6.1 Illustration of a house layout "before adopting biogas," showing a floor plan with three bedrooms, a living room, and a toilet & bathroom. Before they built biogas plat, the living room was located in the middle (*Source* Author)

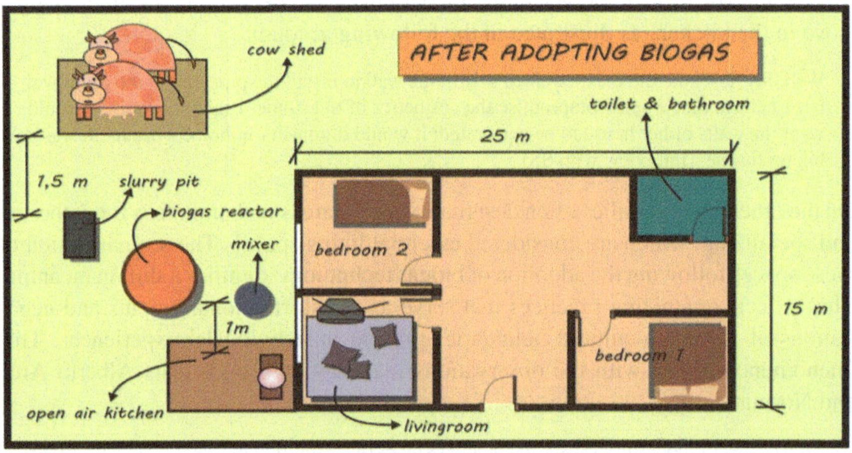

Fig. 6.2 Illustration of a layout "after adopting biogas," showing a residential setup with a biogas system integrated. After adopting biogas, the living room is placed next to open air kitchen (*Source* Author)

Since cows could be sold at the traditional market for profit, instances of cow theft occurred quite frequently, necessitating vigilance among all family members. This design allowed those sleeping in both rooms to observe the cows at any time during the night, enhancing their sense of security. This reconfiguration of space reveals that the family has established a new techno-morality in response to their enthusiasm for technology. The geographical change suggests that high-cost technological innovations, such as biogas, are viewed as shared family assets. They have invested money in constructing the digester, meaning that once the biogas is produced, it becomes their private capital. When "private capital" is established based on accumulated money and energy, the owner's responsibility is to maintain and care for it, ensuring that their efforts and sacrifices in acquiring it do not go to waste. This rationality has driven the restructuring of physical space in this household since the adoption of biogas technology.

This kind of spatial rearrangement aligns with the *nest building* concept proposed by Wilk and Wilhite (1984). This social practice involves altering the material structure, architecture, and usage of space to meet social expectations. In this instance, the homeowner anticipates that biogas technology can be monitored from inside the residence. This nest construction is tied to the symbolic significance of the technological item within the home, rather than solely fulfilling practical needs. The household's arrangement communicates to the outside world their happiness, elegance, and prosperity. The installation of biogas technology, for example, piques the interest and admiration of their neighbors. One neighbor remarked, "Wow, this is really too advanced for me. I probably won't be able to afford it. Your money must be a lot." This indicates that biogas has become a "status marker" for its users (Smaldino et al., 2017). This "social" achievement, as argued by Solomon (2002), is due to the fact that the physical presence of biogas technology is contributive to the leveling of informant's social status. There are more details in his Weberian perspective as follows:

> The construct of status in the present context thus actually refers to the *process of achieving status;* the consumption activities and desires related to the attainment of a certain placement in a social hierarchy by an individual. (Solomon, 2002: 65)

The use of commodities and other items to signal one's social position is known as "status marking," a widespread practice in many cultures (Douglas & Isherwood, 1979). One's social standing is reflected in how they organize their possessions in culturally appropriate ways, distinguishing themselves (or their families) from others. In a sociologist Pierre Bourdieu's term, this is so-called distinction (Bourdieu, 1984).

6.1.3 Traditional and Religious Activities Affecting Biogas Use

Energy consumption practices significantly influence the dynamic use of renewable technologies, and biogas technology is intertwined with locally held religious activities. Religious events entail specific practices that dictate the ways energy is consumed for these purposes. In this context, religious worldviews shape the energy consumption processes in people's everyday lives. Douglas and Isherwood (1979) assert that individuals with religious beliefs exhibit different approaches to managing energy sources. This is evident in the use of biogas within households in this village. For instance, regular religious practices, such as collective Koran recitals, are preserved by households. These events typically take place at the home of the eldest family member, who, in Javanese culture, is viewed as the custodian of cultural preservation and family traditions (Atmawati, 2020; Soepardo, 2018). Since these gatherings often include around 30–40 attendees, substantial amounts of food and beverages must be prepared, requiring cooking time ranging from 2 to 6 h. A household user recognized this limitation of biogas, noting that it cannot operate for more than 8 h a day (Interview with MJ). In contrast, an LPG stove is preferred for longer cooking durations, as it heats up quickly—in just 2 min—and a 20 kg LPG cylinder can sustain continuous cooking for over 6 h (Interview with SN). In light of this situation, household users are able to differentiate and prioritize their energy sources, determining which specific occasions are best suited for biogas use while mobilizing resources effectively for their cooking needs. This illustrates that biogas does not impose constraints; instead, users must make informed decisions about the equipment that best supports their activities during special occasions.

Typically, energy consumption practices in research are linked to demand and resource availability. However, as evidenced in this case, energy consumption cannot be detached from social and religious practices. Wilhite and Lutzenhiser (1999) emphasize that the demand for and volume of energy use are heavily influenced by religious and social customs preserved within the community. This perspective highlights that economic analysis alone cannot resolve the complexities of energy consumption. Furthermore, this activity is gendered, with local women often downplaying their roles and responsibilities in organizing collective cooking during religious events. Women are usually the active participants who take on the majority of household-related activities, including cooking for weddings and religious gatherings. Their control over energy consumption is reflected in their critical choices regarding cooking equipment. One woman recalled:

> The highest decision on what we will cook and how much is nested in our choices. When it comes to household energy spending, we are, of course, much more aware of it than men. During *pengajian* (a communal Koran recitation), we primarily use LPG stoves rather than biogas. It's not that we don't want to use biogas, but for the large amounts of food and beverages we need to prepare, biogas is simply not suitable. It's good for many dishes, but it cannot sustain long cooking times. (Interview with SN)

The narrative above also demonstrates that the use of biogas in the community's religious event exemplifies a technology that is culturally rooted and frequently religiously infused in major areas of economic activity (Coleman, 2022). This author's ethnographic investigation plays vital roles in exploring the relations between religion and economic activity as in the Coleman's (2022) study as follows:

> Ethnographic approaches are shown to be fundamental to understanding the culturally embedded and often religiously inflected character of key areas of economic activity, including gifts, exchange, debt, prosperity and neoliberal entrepreneurship. (Coleman, 2022: 204)

On the other hand, this indicates that women often hold significant power in deciding whether to use biogas or LPG stoves for such activities. However, this situation reveals that the roles of women in the transition to biogas usage cannot be underestimated. In the discourse surrounding consumption, the kitchen becomes a social space for interaction between women who understand the ecological and economic benefits of biogas for their families and those who lack sufficient knowledge about it. This interview highlights the belief that energy consumption is intricately linked to gender dynamics, reflecting women's social positions in spaces designated for consumption (Shove, 2002; Wilhite et al., 2000).

The authors' observations affirmed a tendency where women with a good understanding of biogas felt incapacitated to influence other women, as they believed that such knowledge was only applicable to their own kitchens. Furthermore, differing levels of exposure to technical knowledge regarding biogas made these women more considerate when sharing information with their neighbors. Interestingly, men tended to distance themselves from educating their wives about the biogas cookstove. Even men who were familiar with the technical aspects of biogas appeared less influential when it came to making decisions about whether to use biogas or LPG stoves.

> There is nothing we can do when women occupy the kitchen area. We are very reluctant to engage in cooking activities because we believe we are not meant to work with household appliances. (Interview with MJ)

Additionally, it is clear that, to some extent, women are perceived as having the primary responsibility for collective consumption and are therefore capable of reconfiguring biogas use for communal purposes, as illustrated in this case. This suggests that energy consumption in situations where women have decision-making authority can be seen as an unintended consequence of their social positioning in other areas (Kuehn, 1999). In other words, the realm of communal cooking during religious occasions is a space dominated by women, where they cannot be easily supplanted by men. Although users of biogas are strongly encouraged to rely primarily on their biogas stoves rather than older equipment like LPG stoves, their choices regarding cooking equipment cannot be mandated immediately. This case highlights that the institutional layering involving various actors and their social institutions must be closely examined. At this point, we can advance our understanding that institutional layering directly and indirectly impacts the production, management, and use of energy, while also influencing how societal interactions and interests shape energy consumption practices.

6.2 Conversion

For the household biogas brings advantage and improvement of life. However, since the social setting of families and their needs are diverse, positive impacts may be perceived differently. The conversion of biogas technology into the daily lives of consumers demonstrates that this technology has been included into the domains of everyday life framework (Kossoff, 2011). The following are user stories of how their lives changed after installing biogas.

6.2.1 Feeling of Safety

Operationally, the gas extracted from a mixture of cow dung and water is stored in the upper part of the dome in the biogas reactor. Through a connected pipeline, it is channeled to the biogas stove in the kitchen. Before cooking, the user must rotate the tap controller to allow the gas to flow to the stove. Since the pressure in the biogas system is low, the flame ignites less aggressively compared to an LPG gas stove. LPG, with its high pressure, poses a greater risk of explosion. The prevalence of gas explosions in Indonesia has heightened public concern regarding LPG use. However, users in the village expressed a much better perception of risk associated with biogas; one housewife noted that biogas operates at very low pressure compared to LPG, which alleviates safety concerns.

Moreover, the majority of housewives who transitioned from LPG and kerosene to biogas reported a sense of relief, as they believe biogas is inherently safer due to its low-pressure design. Before adopting biogas, these women often spent longer periods in the kitchen, anxious about potential accidents with their LPG stove (Interviews with SN, NS, and TN). This anxiety limited their ability to engage in other household tasks, but after switching to biogas, they found they could multitask—watering plants, helping children with their studies, reading the newspaper, or enjoying a book in the living room while waiting for food to cook. The shift to biogas has enabled women to step away from the stove, thus enhancing their participation in family life and allowing for quality time with their children. This phenomenon illustrates how users and biogas, as a technological product, interact to create "a set of relationships that give them their shape" (Law, 2000: 5). The relationship that emerges from this transition shows that women and biogas develop mutual associations: women experience better social interactions with family members, while biogas provides sustainable energy for the household. This successful co-productive transformation comes to realization because "user's degree of satisfaction with the artifact as it is, the availability of symbolic and material resources to undertake transformative actions" (Boczkowski, 1999: 91).

The perception of zero risk is further evidenced by the reduced fear of allowing children to operate the biogas stove. During an interview at SK's house, he asked his 10-year-old daughter to boil water for my coffee. He justified this by stating that the stove is relatively safe for children, provided parents teach them how to operate it properly and ensure they are capable of doing so independently (Interview with SK). This dynamic may reflect a broader participation in the creation of technology. Involving children in using the biogas stove indicates that the technology has been designed with user customization in mind, suggesting that users play an active role in the development of such technologies (Brady et al., 1992).

6.2.2 No More Reliance on Firewood and Environmental Consciousness

Following Lie and Sørensen (1996) and Berker et al. (2005), the integration of biogas technology fosters the co-production of meaning, highlighting the long-term benefits associated with its use. An illustrative example can be seen in the experiences of users who have transitioned away from firewood for household cooking. Although firewood is an abundant resource in the village, it becomes damp and unusable during the rainy season, compelling households that rely on it to seek alternative solutions. Biogas has emerged as an effective option, enabling users to cook without concern for rain, resulting in savings of approximately IDR 50,000–100,000 per month previously spent on purchasing firewood from local markets (Interview with SD). Households that relied on firewood consumed more than 20 kg each month, incurring significant costs and dependence over the years. They also faced health risks associated with ash, which can be harmful to lung health. While no formal medical studies have been conducted to evaluate specific health issues, several housewives reported frequent coughing while cooking with firewood. Cooking with this traditional fuel source has proven dangerous, particularly for women. Although the link between deforestation and minor flooding in rural areas is not widely recognized among users, some have begun to understand this pressing issue.

In contrast, biogas use has raised awareness of various environmental problems, such as the harmful effects of methane gas on the atmosphere. This newfound understanding emerged through socialization and interaction with diverse stakeholders involved in the biogas project. One user articulated this concern:

> Biogas has brought new knowledge that we, as villagers, might not have realized, such as the possibility of flooding due to the rapid decline of trees that used to protect our environment. By using biogas, we are practicing moral values and demonstrating our responsibility toward nature. This is a small step that can lead to broader change. (Interview with SN)

Transitioning to biogas has positively impacted households, reducing health risks associated with firewood combustion that can place women in vulnerable situations, while also altering perceptions of environmental sustainability. This practice reflects the adaptability, improvisation, and willingness to experiment exhibited by biogas users (Evans, 2011).

6.2.3 Savings for Education Funds and the Shift of Gender Role in Family

Children's education is a fundamental aspect of human development, and the use of biogas allows households to save money for financing their children's schooling. For families not engaged in farming, these monthly savings are often allocated for educational expenses, as illustrated by SS's experience. After transitioning to biogas, his family can save up to IDR 150,000, which they use to buy books for his daughter, who is studying at the university in Yogyakarta (Interview with SS). This amount is equivalent to their monthly expenditure on an LPG gas cylinder needed to cook for five family members (two parents and three daughters). Because the biogas system is consistently maintained, they never experience a shortage of gas. SS expressed satisfaction, noting that before switching to biogas, he had never bathed with warm water because doing so would significantly increase their LPG expenses. With biogas, he can now take warm baths almost every day, a luxury he was previously unable to afford.

For households engaged in farming, biogas also helps save on monthly expenses, particularly during the harvesting season from June to August. While savings vary among households, many report saving around IDR 200,000 during this period, as they need to provide food and beverages for five laborers assisting in the cultivation of their land. A housewife whose husband works in the paddy fields shared how biogas has transformed their lives. She emphasized that after integrating biogas into their cooking, she could pay both of her children's tuition fees (Interview with TA). Given the rising costs of secondary education in Indonesia, biogas has been tremendously beneficial for her family.

Moreover, the introduction of biogas has positively impacted the division of child-rearing responsibilities between her and her husband. Her husband now prefers to wake up an hour earlier than she does to boil water and prepare breakfast for their children, knowing she must leave for school early at 7 a.m. His commitment to checking the biogas plant each morning has increased his sense of responsibility in supporting his wife with household duties. TG, the husband, is now very happy and willing to care for his daughter, something he hadn't done before. He has realized that assisting his wife with household tasks helps strengthen his emotional bond with their little daughter.

6.2.4 From Technology to Social Status and Political Opportunity

As defined by the domestication of technology approach, conversion pertains to how households engage with the outside world and the ways in which household members align technology with societal values and the socio-cultural opportunities it presents after being introduced to the community. Moreover, the consumption of household technology plays a role in shaping individuals' identities through their relationships with those around them. From this conceptual framework, examining how biogas technology produces *socio-cultural opportunities* for its users becomes essential (Glaveanu et al., 2020; Lemay et al., 2019; Sheng et al., 2019).

Non-users of biogas often perceive it as a modern good, symbolizing material wealth and social status. This perception likely arises because the construction of biogas systems requires a significant financial investment. From the perspective of non-users, biogas adoption represents a shift in social status, suggesting that households with biogas are viewed as more affluent than their counterparts without it. When asked about the impacts of biogas on household modernity, many non-user informants indicated that biogas has signified a distinction in social standing, elevating the perceived status of those who utilize it. In Javanese culture, cows embody cultural significance and symbolize high status, so families that adopt biogas—often alongside livestock—are viewed as socially superior. Households with biogas are frequently regarded as wealthier because they can allocate funds toward technological needs, while others may struggle to cover educational expenses for their children. This dynamic indirectly creates a "status hierarchy" among villagers (Kim, 2018). The perception of biogas as a domesticated technology associated with innovation and financial capability reinforces this hierarchy (Winther & Bell, 2018).

In fact, such shift in social status post-biogas adoption opens new avenues for households to access village information and collaborate with external organizations, such as government bodies. Households that adopt biogas gain equal standing in the community, distinguishing them from their peers. For instance, SK, a biogas user, felt empowered to propose a development project to the village chief, a move he might not have considered before adopting the technology. His newfound confidence even led him to run for a position in the Village Parliament in 2011 (Interview with MJ). This illustrates a compelling interpretation of how possessing renewable energy technology intertwines with political opportunities. SK's experience suggests that biogas adoption fosters a collective emergence of political acceptance, allowing users to embrace new identities, initiatives, and possibilities. This flexibility represents a new logic of differentiation (Eckhardt & Bardhi, 2019).

On the other hand, biogas technology also allows users to embody an image of modernity, as it involves innovation and advanced knowledge for its construction and use. The adoption of biogas facilitates new social interactions and elevates social status, creating emergent relationships between users and their community. As SB, the village secretary, noted, the impacts of biogas technology on social dynamics are profound and transformative:

> They are no longer laypersons in this village, as they once were. The adoption of biogas technology marks a significant change, as its cost and the need for technical proficiency in construction and maintenance make it inaccessible to everyone. Those who do possess biogas systems are viewed differently; they are seen as modern, equipped with a technology that is rare and associated with a higher social status compared to other villagers. Additionally, biogas users tend to attend community meetings more frequently than before, illustrating how technology fosters greater social interaction among its users.

Subjective satisfaction is further enhanced by the perception that biogas technology provides community benefits. Users feel that their biogas systems are not only useful for personal use but also fulfilling in a communal context. Local leaders recognize this engagement, playing a crucial role in shifting the purpose of biogas from individual households to communal welfare. This process can be described as "social engagement" (Bixter et al., 2018: 179). One respondent emphasized this transformation:

> My biogas is not just for our family. The benefits of organic bio-slurry have attracted many neighbors, who often come to our bio-slurry hut to collect it for feeding their catfish in their own ponds. My husband is frequently invited by the village chief to train other local residents. The point is that technology impacts and shares blessings with everyone around us. Thank God we have biogas.

This interview highlights that biogas technology serves as an artifact that fosters "social benefits" (Cui & Sun, 2019: 16), stemming from the shared advantages of bio-slurry as an organic fertilizer for the community. Users experience a growing sense of mutual sharing toward their neighbors, bolstered by the technology's role in sustaining social welfare (Kruks-Wisner, 2018; Prawira et al., 2021; Taylor-Gooby, 2019). The rising cost of fish feed has become a pressing issue among agricultural farmers in some areas, drawing collective concern for solutions. As AG expressed:

> I discern that biogas embodies a humanistic value, such as solidarity, perceived by its users. Having biogas creates a 'spill-over' effect to your neighbors and community, stemming from the benefits you derive from using it. Since I adopted biogas, many people have asked me to share the bio-residue for feeding their fish, especially as fish food has become increasingly expensive. With the technology you possess, it's essential to engage more closely with your community.

The discussion above reveals how biogas technology has become ingrained in the daily lives of users. Its integration into local interactions is largely attributed to the high level of acceptance from non-users, who recognize the significant advantages it offers. Non-users play a crucial role in making biogas technology socially relevant within the community. Without their acknowledgment, biogas would remain merely a functional artifact, devoid of deeper meaning. This demonstrates that biogas has catalyzed new social interactions, giving rise to a new culture rooted in the social significance of the technology (Kroeber & Kluckhohn, 1952).

Technological conversion is vital to the concept of transition, facilitating the continuous use of biogas by its users. Transition is essential for creating more sustainable futures, and design plays a critical role in this process. Modifying technical components can also be considered a form of technological conversion, shaped by users' engagement in social practices, such as religious events. During fieldwork, I observed that one user added an extra biogas stove in their kitchen. This was primarily due to an anticipated increase in family food consumption and the desire to avoid using LPG for daily cooking. The additional stove was connected via a branched pipeline to ensure equal distribution of biogas. This innovation also included aesthetic upgrades to the kitchen, such as ceramic flooring (see Fig. 6.3). Regarding these modifications, SN explained,

> I decided to branch the biogas pipeline in the kitchen because I wanted an extra stove. With our food consumption expected to rise, especially during Ramadan, I need to cook a variety of dishes for breaking the fast. I would also feel embarrassed if the kitchen wasn't decorated with ceramic flooring because when my daughter-in-law's family visits, they might be put off by a dirty and messy kitchen.

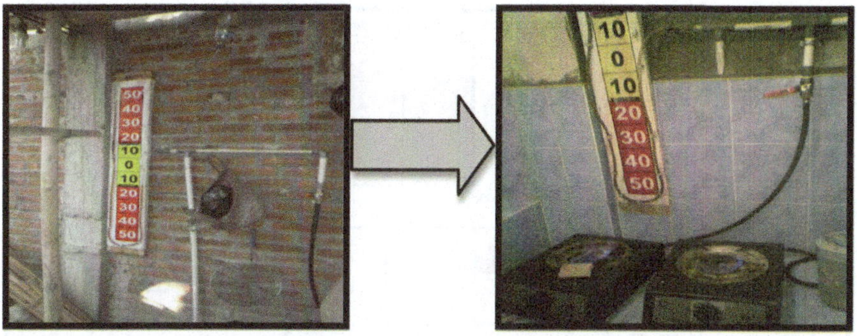

Fig. 6.3 The new kitchen arrangement after installing biogas. The left image displays an unfinished room with exposed brick walls and a manometer on the wall reading water indicator to inform the users of the gas volume available. The higher the water indicator line, the higher the gas volume users can use (*Source* Author)

The statement above highlights that adjustments to the biogas infrastructure are driven not only by functional needs but also by aesthetic considerations. Cleanliness and neatness are tied to the expectations of others, rather than just technical requirements. This suggests that the decision to modify the technology and domestic space is influenced by social judgments, reflecting the concept of social construction. Consequently, technical maintenance and user innovation foster "functions originally embedded in localized relationships" (Smith, 1987: 6).

In line with Bryan Pfaffenberger's (1992) concept of technical adjustments, the users make changes to integrate a new technology or manufacturing method into their home environment. These adjustments go beyond the technical realm and involve adapting the technology to fit social, spatial, and cultural dynamics within the household and community. Pfaffenberger defines these changes as strategies users employ to mitigate or leverage the social impacts brought about by new technologies. In the case of biogas, the spatial positioning of the biogas plant allows for inclusive interactions between users and non-users alike. This is particularly evident among users who hold leadership positions within their community. For example, a user with leadership responsibilities redesigned his biogas space to "engage in strategies that try to compensate for the social esteem and social power that the technology has caused" (Pfaffenberger, 1992: 286). The sketch displayed in Fig. 6.4 illustrates the strategic placement of the biogas plant in relation to the community hall at TG's house.

The leadership role of TG, the head of the sub-village, places him at the center of both social and technological transformations within his community. His monthly community meetings, a regular event attended by household leaders, offer a platform to discuss pressing societal issues. After these meetings, TG encourages attendees to view his biogas plant, located just five meters from the community hall. This interaction not only familiarizes the community with the biogas technology

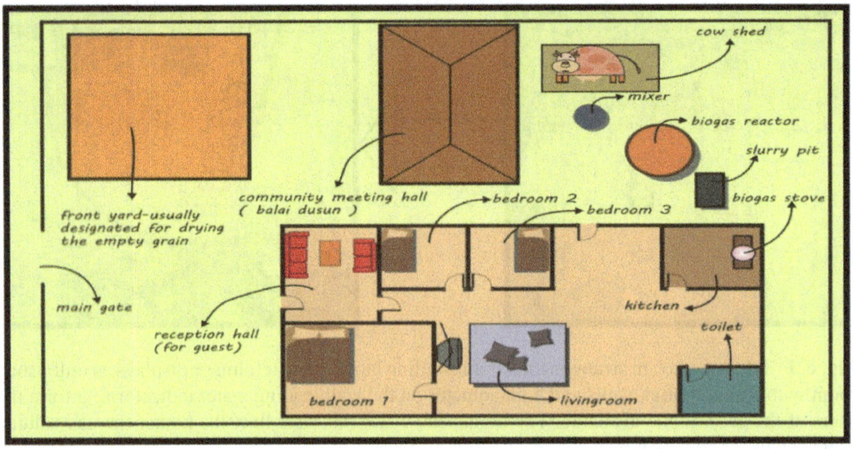

Fig. 6.4 The position of biogas plant next to community hall. Illustration of a sustainable living compound layout, featuring a main house with labeled rooms including bedrooms, kitchen, living room, and toilet. Adjacent structures include a community meeting hall, a cow shed with a biogas system comprising a mixer, biogas reactor, and stove connected to a slurry pit. The front yard is noted for drying grain, and there's a reception hall for guests. The design emphasizes eco-friendly living with integrated animal and biogas facilities (*Source* Author)

but also imparts a deeper cultural meaning, as the technology symbolizes modernity and technological progress within the village. In the eyes of the community, TG's biogas plant represents both a technological advancement and a form of social status, highlighting his position as a power holder. However, this social recognition brings positive outcomes, opening up opportunities for TG to collaborate with external experts from Indonesian universities to enhance the organic paddy fields in his sub-village (Interview with TG and Interview with SB). This illustrates how the relationship between biogas technology and its users extends beyond mere functionality. It integrates into various domains of everyday life, including political, expertise, and cultural fields, as suggested by Gardiner (2000).

The technical changes and modifications users have made to their biogas systems stem from their local knowledge, ensuring that the technology aligns with their everyday farming practices and long-standing local traditions. Grin (2010) support this approach, emphasizing the significance of local knowledge in driving technological transitions:

> To acknowledge the importance of local knowledge which is needed to apply change in specific settings [...] a transition agenda should also be based on the local knowledge of consumers and local communities. (p. 331)

6.3 Conclusion

The findings highlight how biogas technology has been seamlessly integrated into the daily routines of users, emphasizing the role of social actors and institutional co-shaping in facilitating this integration. The process of biogas incorporation extends beyond its technical usage and involves cultural and social negotiation. This aligns with Bell's (2001) concept that technology organizes human experience through instrumental means, while Berker et al. (2006) emphasize how the combination of technology, culture, and its placement in both public and private spaces can create social power, elevating the user's position within their community. The political and social implications of biogas technology are significant. As seen in the case of biogas users gaining a political bargaining position in the village, technology can enable social mobility and prestige. Wilhite and Lutzenhiser (1999) explore this dynamic, noting that ownership and usage of technology can reflect social status and trigger upward mobility. This reinforces the notion that technological artifacts, like biogas, contribute to shaping one's social standing within the community.

Incorporating biogas also involves reconciling its use with pre-existing cultural practices and consumption habits. If users cannot align biogas usage with their established routines, there is a risk of reverting to fossil-fuel-based practices, which are environmentally unsustainable. This highlights the importance of cultural adaptation in ensuring long-term sustainable energy use. Winther (2008) extends this idea by connecting material possession with cultural capital. Displaying biogas technology becomes a form of "cultural text," representing not just a technological object but a symbol of social power and personal advantage. The biogas plant becomes part of the household's symbolic capital, showcasing both economic means and social taste, akin to Bourdieu's theory of cultural capital.

Ultimately, the findings indicate that biogas technology has transcended its function as a mere energy source and has been incorporated into broader socio-political and cultural contexts. The domestication framework of incorporation and conversion illustrates how biogas has become transformative, impacting users' lives and reshaping their social interactions. The technology's acceptance by non-users further solidifies its role as a socially meaningful artifact, influencing both the personal and collective spheres of life.

References

Arce, A., & Long, N. (2003). Reconfiguring modernity and development from an anthropological perspective. In *Anthropology, development and modernities* (pp. 15–44). Routledge.

Atmawati, D. (2020). *The shift of proper names among Javanese Society*. In 4th International Conference on Language, Literature, Culture, and Education (ICOLLITE 2020) (pp. 202–207). Atlantis Press.

Bell, D. (2001). *The future of technology*. Pelanduk Publications.

Bensch, G., & Peters, J. (2015). The intensive margin of technology adoption—Experimental evidence on improved cooking stoves in rural Senegal. *Journal of Health Economics, 42*, 44–63. https://doi.org/10.1016/j.jhealeco.2015.03.006

Berker, T., Hartmann, M., Punie, Y., & Ward, K. (2005). *Domestication of media and technology*. Open University Press.

Berker, T., Hartmann, M., Punie, Y., & Ward, K. (2006). Introduction. In T. Berker, M. Hartmann, Y. Punie, & K. Ward (Eds.), *Domestication of media and technology*. Open University Press.

Bixter, M. T., Blocker, K. A., & Rogers, W. A. (2018). Enhancing social engagement of older adults through technology. *Aging, Technology and Health*, 179–214. https://doi.org/10.1016/b978-0-12-811272-4.00008-7

Boczkowski, P. J. (1999). Mutual shaping of users and technologies in a national virtual community. *Journal of Communication, 49*(2), 86–108.

Bourdieu, P. (1963). La société traditionnelle: Attitude à l'égard du temps et conduite économique. *Sociologie Du Travail, 5*(1), 24–44. https://doi.org/10.3406/sotra.1963.1127

Bourdieu, P. (1984). *Distinction*. Routledge.

Bourdieu, P. (1986). The forms of capital. In J. G. Richardson (Ed.), *Handbook of theory and research for the sociology of education* (pp. 241–258). Greenwood.

Brady, T., Tierney, M., & Williams, R. (1992). The commodification of industry applications software. *Industrial and Corporate Change, 1*(3), 489–514. https://doi.org/10.1093/icc/1.3.489

Coleman, S. (2022). Ritual, rationality and intersections between economy and religion. In *A handbook of economic anthropology* (pp. 204–215). Edward Elgar.

Cui, Y., & Sun, Y. (2019). Social benefit of urban infrastructure: An empirical analysis of four Chinese autonomous municipalities. *Utilities Policy, 58*, 16–26. https://doi.org/10.1016/j.jup.2019.03.001

Douglas, M., & Isherwood, B. (1979). *The world of goods*. Allen Lane.

Eckhardt, G. M., & Bardhi, F. (2019). New dynamics of social status and distinction. *Marketing Theory, 20*(1), 85–102. https://doi.org/10.1177/1470593119856650

Evans, D. (2011). Consuming conventions: Sustainable consumption, ecological citizenship and the worlds of worth. *Journal of Rural Studies, 27*(2), 109–115. https://doi.org/10.1016/j.jrurstud.2011.02.002

Fraune, C. (2015). Gender matters: Women, renewable energy, and citizen participation in Germany. *Energy Research & Social Science, 7*, 55–65. https://doi.org/10.1016/j.erss.2015.02.005

Fritzsche, P. (2001). Specters of history: On nostalgia, exile, and modernity. *The American Historical Review, 106*(5), 1587–1618. https://doi.org/10.1086/ahr/106.5.1587

Gardiner, M. E. (2000). *Critiques of everyday life*. Routledge.

Giddens, A. (1991). *Modernity and self-identity: Self and society in the late modern age*. Polity Press.

Glaveanu, V. P., Ness, I. J., & de Saint Laurent, C. (2020). Creativity, learning and technology: Opportunities, challenges and new horizons. *Creativity Research Journal, 32*(1), 1–3. https://doi.org/10.1080/10400419.2020.1712167

Gonda, N. (2016). Climate change, "technology" and gender: "Adapting women" to climate change with cooking stoves and water reservoirs. *Gender, Technology and Development, 20*(2), 149–168. https://doi.org/10.1177/0971852416639786

Grin, J. (2010). Understanding transitions from a governance perspective. In *Transitions to sustainable development*. Routledge.

Kim, H. (2018). The mutual constitution of social media use and status hierarchies in global organizing. *Management Communication Quarterly, 32*(4), 471–503. https://doi.org/10.1177/0893318918779135

Kossoff, G. (2011). Holism and the reconstitution of everyday life: A framework for transition to a sustainable society. In S. Harding (Ed.), *Grow small, think beautiful*. Floris Books.

Kroeber, A., & Kluckhohn, C. (1952). *Culture: A critical review of concepts and definitions*. Harvard University Press.

Kruks-Wisner, G. (2018). *Claiming the state: Active citizenship and social welfare in rural India*. Cambridge University Press.

Kuehn, S. (1999). *Livsstilens betydning for energiforbruget*. Københavns Universitet, Sociologisk Institut.

Kumar, P., & Mehta, S. (2016). Poverty, gender, and empowerment in sustained adoption of cleaner cooking systems: Making the case for refined measurement. *Energy Research & Social Science, 19*, 48–52. https://doi.org/10.1016/j.erss.2016.05.018

Law, J. (2000). *Economics as interference*. Centre for Science Studies and Department of Sociology.

Lemay, D. J., Doleck, T., & Bazelais, P. (2019). Context and technology use: Opportunities and challenges of the situated perspective in technology acceptance research. *British Journal of Educational Technology, 50*(5), 2450–2465. https://doi.org/10.1111/bjet.12859

Lie, M., & Sørensen, K. (1996). Making technology our own. In *Domesticating technology into everyday life*. Scandinavian University Press.

Malhotra, P., Neudoerffer, R. C., & Dutta, S. (2004). A participatory process for designing cooking energy programmes with women. *Biomass and Bioenergy, 26*(2), 147–169. https://doi.org/10.1016/s0961-9534(03)00083-7

Musharbash, Y. (2007). Boredom, time, and modernity: An example from Aboriginal Australia. *American Anthropologist, 109*(2), 307–317. https://doi.org/10.1525/aa.2007.109.2.307

Pakulski, J. (1997). Cultural citizenship. *Citizenship Studies, 1*(1), 73–86. https://doi.org/10.1080/13621029708420648

Pfaffenberger, B. (1992). Technological dramas. *Science, Technology, & Human Values, 17*(3), 282–312. https://doi.org/10.1177/016224399201700302

Popp, D., Hascic, I., & Medhi, N. (2011). Technology and the diffusion of renewable energy. *Energy Economics, 33*(4), 648–662. https://doi.org/10.1016/j.eneco.2010.08.007

Prawira, W. R., Maulida, H., & Achmad, W. (2021). Narrating the implementation of social welfare community program. *Rigeo, 11*(5), 228–235. https://rigeo.org/menu-script/index.php/rigeo/article/view/763

Risman, B. J., & Davis, G. (2013). From sex roles to gender structure. *Current Sociology, 61*(5–6), 733–755. https://doi.org/10.1177/0011392113479315

Rizzi, F., van Eck, N. J., & Frey, M. (2014). The production of scientific knowledge on renewable energies: Worldwide trends, dynamics and challenges and implications for management. *Renewable Energy, 62*, 657–671. https://doi.org/10.1016/j.renene.2013.08.030

Schmid, B., & Smith, T. S. (2020). Social transformation and postcapitalist possibility: Emerging dialogues between practice theory and diverse economies. *Progress in Human Geography, 45*(2), 253–275. https://doi.org/10.1177/0309132520905642

Sheng, J., Amankwah-Amoah, J., & Wang, X. (2019). Technology in the 21st century: New challenges and opportunities. *Technological Forecasting and Social Change, 143*, 321–335. https://doi.org/10.1016/j.techfore.2018.06.009

Shove, E. (2002). *Comfort, cleanliness and convenience. The social organization of normality*. Palgrave Macmillan.

Shove, E. (2012). Comfort and convenience: Temporality and practice. In F. Trentmann (Ed.), *The Oxford handbook of the history of consumption*. Oxford University Press.

Smaldino, P. E., Janssen, M. A., Hillis, V., & Bednar, J. (2017). Adoption as a social marker: Innovation diffusion with outgroup aversion. *The Journal of Mathematical Sociology, 41*(1), 26–45.

Smith, D. (1987). *The everyday world as problematic: A feminist sociology*. Northeastern University Press.

Soepardo, S. (2018). Address term in a family of Javanese Priyayi. *Leksika: Jurnal Bahasa, Sastra Dan Pengajarannya, 1*(2). http://jurnalnasional.ump.ac.id/index.php/LEKSIKA/article/view/2259

Solomon, M. R. (2002). The value of status and the status of value. In *Consumer value* (pp. 79–100). Routledge.

Swedberg, R. (2011). The economic sociologies of Pierre Bourdieu. *Cultural Sociology, 5*(1), 67–82. https://doi.org/10.1177/1749975510389712
Taylor-Gooby, P. (2019). *Social change, social welfare and social science*. University of Toronto Press.
Uchenna, E., & Oluwabunmi, A. O. (2020). Cooking technology and female labor market outcomes in sub-Saharan Africa. *African Development Review, 32*(4), 661–672. https://doi.org/10.1111/1467-8268.12468
Wallenborn, G., & Wilhite, H. (2014). Rethinking embodied knowledge and household consumption. *Energy Research & Social Science, 1*, 56–64. https://doi.org/10.1016/j.erss.2014.03.009
Wilhite, H., & Lutzenhiser, L. (1999). Social loading and sustainable consumption. In E. J. Arnould & L. M. Scott (Eds.), *Association for consumer research* (vol. 26, pp. 281–287). .
Wilhite, H., Shove, E., Lutzenhiser, L., & Kempton, W. (2000). *Twenty years of energy demand management: We know more about individual behaviour but how much do we really know about demand?* Proceedings of ACEEE 2000 Summer Study.
Wilk, R., & Wilhite, H. (1984). Household energy decision making. In W. Kempton & B. Morrison (Eds.), *Families and energy: Coping with uncertainty*. Michigan State University.
Winther, T. (2008). *The impact of electricity: Development, desires and dilemmas*. Berghahn Books.
Winther, T., & Bell, S. (2018). Domesticating in home displays in selected British and Norwegian households. *Science & Technology Studies, 31*(2), 19–38. https://doi.org/10.23987/sts.56791

Chapter 7
Conclusion

Abstract The domestication of biogas technology involves the dynamic interplay of socio-cultural resources and social skills from various stakeholders, including users, the target community, the government, and constructors. Biogas technology, originally designed to assist with household tasks, has evolved into a social artifact as its functionality is reinterpreted and culturally adapted. This process transforms the technology from a mere tool into a medium through which social and cultural practices are expressed. The study highlights that adopting biogas technology is closely tied to existing energy consumption habits within households. As families incorporate biogas into their routines, they are required to adjust their consumption patterns to reflect pre-existing social practices. This process of adaptation illustrates the complexity of technological transition, where the shift to biogas as the primary cooking technology is often marked by instability. Households must negotiate between new technological practices and ingrained cultural behaviors, making the initial period of biogas adoption a time of fluctuating usage and gradual adjustment. This phenomenon underscores that the successful adoption of new technology involves more than its technical features; it requires a harmonization of the technology with cultural and social norms. Biogas technology, therefore, is not only an energy solution but also a medium through which households negotiate and adapt their consumption practices, reflecting broader cultural and social dynamics.

Keywords Domestication · Biogas · Social practices · Adoption

This book primarily examines how biogas technology is integrated into households by users. Using the domestication of technology as a theoretical framework—consisting of appropriation, objectification, incorporation, and conversion—the arrival of biogas technology into the home requires various resources and prompts significant changes in family dynamics. These changes emerge as a result of the co-shaping process between the technology and the users.

The market acts as a space where potential users and technology providers interact. However, the technology becomes familiar to users not just through direct contact with providers but also through social networks and organizational affiliations. This

M. Alam, *The Social Life of Biogas*, https://doi.org/10.1007/978-981-96-3300-5_7

study uniquely highlights a case where users and biogas organizing institutions were closely affiliated even before the technology was introduced. These users were already involved in social activities organized by this institution, which had grassroots networks across Bantul, Indonesia, centered around an organic farming platform. From the previous chapters' discussions and findings, it becomes evident that the successful adoption of biogas at the household level is largely influenced by interpersonal networks. These networks are critical spaces where trust is established, and information about the technology is reliably shared. Before adopting biogas, households often seek guidance from individuals within their close social circles. Kinship serves as an emotional mediator, facilitating the exchange of information and problem-solving strategies (Meng et al., 2023). This process helps potential users feel more confident and at ease.

Once the biogas system is installed, users who encounter operational challenges often turn to those with whom they share kinship for assistance. In terms of financing, the money spent on construction materials influences the household's long-term commitment to the technology. Moreover, the flexibility to reuse materials from existing house structures reduces costs. Strong social capital, in the form of cooperation among villagers, remains a crucial element. As Hasan et al. (2022) argue, mutual assistance is particularly meaningful during the initial feeding phase to activate the biogas plant.

The users are not passive adopters; they are innovative. Even before adopting biogas, they often have their own ideas, knowledge, and expectations about reactor designs that will meet their specific needs. This aligns with Kanda and Hjelm's (2021) study, which highlights the co-learning that occurs between the Community-based Project Operators (CPOs) and the users to ensure smooth biogas operations. As users incorporate biogas into their daily routines, they must acquire new knowledge and skills to operate the system effectively. Once they understand how the system functions, they adjust its operations to fit their daily routines. In some cases, users have innovated by modifying installation designs to make biogas operations more convenient (Wang et al., 2020).

Key factors in the successful adoption of biogas technology can offer valuable insights and should be considered in policy decisions. Engaging relevant stakeholders is crucial to making biogas feasible for grassroots communities. Jalas et al. (2014) emphasized the importance of more robust engagement with citizens, consumers, and other social actors within the specific configurations of biogas use. The proximity of biogas users in Sri Hardono village fosters an environment for the exchange of technical skills. After biogas construction, users typically receive brief instructions from the CPO on operating the biogas system and performing regular monitoring. During the first month, new users often encountered minor issues, such as the biogas stove failing to produce a blue flame. In response, experienced users observed and shared strategies to help new users resolve such issues. This interaction highlights the application of technological framing as a means of sustaining the operation of biogas technology (Bijker, 2001).

Given the rural context in Indonesia, it is essential that information and promotions are delivered in local languages, involving village stakeholders, such as chiefs and sub-village leaders. Encouraging potential users to contribute to the construction materials rather than presenting a fixed total cost for each biogas plant can help maintain interest. The author's field experience reveals that when Community Project Organizations (CPOs) present the full cost upfront, potential users' enthusiasm may diminish. Post-construction training is equally critical to ensure all household users have the necessary knowledge and skills to sustain biogas operations (Tembo et al., 2023). This research further suggests that users should be viewed as active participants, equipped with prior knowledge and information. During the construction phase and beyond, their involvement in designing, building, and maintaining the plant can foster a learning process that helps them internalize new insights on biogas maintenance. Biogas technology should also promote the inclusion of women at all stages, from construction to daily operations. In the field, we observed that men typically handle daily operations and maintenance tasks. However, ensuring that biogas technology remains in use long-term requires engaging women and providing them with the skills to manage operational schedules for their families.

This research employs the concept of the domestication of technology to explore how biogas is integrated into household routines. The domestication process, which includes appropriation, objectification, and conversion, helps examine how users and technological artifacts interact and are socially co-constructed within households. The research also highlights how changing perceptions of technology can impact households, evident in the benefits gained from its use. For example, the reduction in perceived risks associated with cooking activities after switching to biogas is a clear benefit. In this process, users create new meanings for their lives by comparing the old and new cooking technologies. Once they perceive that biogas is safer, they make it a routine part of their lives. This aligns with Selwyn's (2003) idea that users understand the benefits of technology through lived experience and evolving values. Applying the four concepts of domestication to biogas adoption, the author argues that the process of integrating artifacts into daily life is not always linear. For instance, one household (referred to as SS) recognized the benefits of biogas only after extensive online research and observing a government-installed biogas plant in their village. This exposure led them to consider which biogas model best suited their needs. Lehtonen's (2003) work supports this, suggesting that appropriation, objectification, incorporation, and conversion are not always orderly steps in the domestication process.

The knowledge users possess about biogas and the socio-cultural values surrounding its use play a critical role in the domestication process (Mackay & Gillespie, 1992; Ward, 2006). Sørensen (2006) adds that cognitive processes and the construction of meaning influence the continuous use of technology. From the domestication theory, the author learns that biogas technology, as an artifact, can serve as a lens through which users interpret the world. This interpretation cannot be separated from the nature of the technology itself, a concept Ihde refers to as "multistability." This idea highlights that technology can be appropriated in various frameworks of interpretation and practice, leading to diverse mediating roles (Verbeek & Vermaas, 2010: 169).

Using this case study, the research delves deeply into households' experiences with biogas technology in everyday life, offering tools to interpret the social meanings embedded in daily interactions in the field (Flichy, 2007). Semi-structured interviews serve as a key method, allowing the research to remain focused on the objectives established before fieldwork commenced. One of the main advantages of qualitative inquiry is its ability to break down barriers between the researcher, informants, and other participants. This method provides access to sensitive information that might otherwise be politically confidential, often revealed through informal conversations, frequent visits, and participation in community events hosted by informants.

In general, the social construction of technology (SCOT) theory requires the author to examine three levels of analysis: (i) sociological construction, which demonstrates interpretative flexibility, (ii) the social construction of technological artifacts, and (iii) the explanation of this construction process in terms of the technological frame of relevant social groups (Bijker, 2009: 10). I acknowledge that this research has not deeply explored biogas technology, as there remain many unobserved aspects within the household. To address this gap, it is vital to apply an ethnographic approach, potentially involving a larger sample of households. Two key concepts in SCOT—technological frame and interpretative flexibility—are crucial in analyzing how biogas usage practices co-evolve. First, interpretative flexibility allows us to investigate how users innovate. For example, some users modify their cowsheds or add quills to the mixture of cow dung and water. Innovation, in this context, depends on users' capacity to adapt the technology to their specific needs. For those who view quills as waste, biogas becomes a means of decomposition. Conversely, users with unchanged cowshed structures innovate by integrating mixers into their cowsheds for greater convenience and comfort. These innovations emerge from users' knowledge and cognitive frameworks, aligning with Wajcman's (2004) argument that user capacity is a crucial dimension of interpretative flexibility. In this research, the domestication of technology is incorporated into the social construction of technology (SCOT) framework to understand how biogas technology is integrated into household routines. By employing the concepts of appropriation, objectification, and conversion, it becomes highly relevant to examine how users and technological artifacts interact and are socially co-constructed within the household sphere. Furthermore, the discussion sheds light on how changes in the value of technology can impact households, particularly through the benefits gained from its use. For example, the reduced perception of risk associated with cooking after switching to a biogas stove is an obvious illustration. In this process, users create new meanings for their lives, comparing the old cooking methods with biogas. Once users perceive that the new technology provides greater safety, they are more likely to incorporate it into their daily routines. This aligns with Selwyn's (2003) assertion that users can sense the benefits of technology through the reflection of lived experience and changing values.

When contextualizing the four concepts of domestication in the adoption of biogas technology, the author proposes that the incorporation of artifacts does not always follow an orderly process. For instance, one user recognized the benefits of biogas after conducting extensive research online and observing a government-installed biogas plant in his village. This exposure generated new cognitive considerations about which biogas model he should choose. This argument is supported by Lehtonen (2003), who notes that appropriation, objectification, incorporation, and conversion do not always follow a linear process in the domestication of technology. From this perspective, users' knowledge of biogas, along with their socio-cultural values, plays a crucial role in the domestication process (Ward, 2006). Supporting this, Sørensen (2006) argues that cognitive processes and the construction of meaning shape the continuous use of technology.

According to Hewitt et al. (2022), the domestication of biogas technology involves not only socio-cultural resources but also the social competencies displayed by users. Biogas is not merely a tool for household tasks; as Chodkowska-Miszczuk (2021) argues, it becomes a social artifact when meaning-making leads to the functional use of the technology. Moreover, the cultural use of this technology helps households manage it properly on a daily basis. Dobers (2019) highlights that within the family space, domestication also involves changes in the household's physical environment. As biogas technology enters the home, co-shaping activities between the artifact and the users emerge, reinforcing the idea that both the artifact and the users influence each other. This mutual influence leads to a new worldview (Cavicchi, 2016). This transformation is observed in how households restructure their living spaces to integrate biogas technology as part of their routines and ownership. Additionally, the domestication of biogas technology is closely tied to energy consumption practices within the family. In appropriating biogas, households must negotiate new consumption patterns while preserving long-established social practices. As a result, the transition period is often marked by an unsettled usage of biogas technology as the primary cooking method. It is important to note that the way users value newly adopted technology is influenced significantly by social institutions, socio-cultural practices, and broader networks (Sime, 2020). Contrary to common assumptions, economic conditions do not appear to significantly affect the sustainable integration of biogas technology into the household's societal life (Forbord & Hansen, 2020; Ismail et al., 2021).

From domestication theory, we learn that biogas technology, as an artifact, can serve as a lens through which users perceive and interpret the world, and this interpretation is inseparable from the nature of the technology itself. Ihde refers to this concept as "multistability," explaining that "human beings can appropriate technologies in different frameworks of interpretation and use practices, which can result in various mediating roles" (Verbeek & Vermaas, 2010: 169).

While this approach has its challenges, such as determining the reliability of materials for analysis, the use of data triangulation helped overcome these hurdles. Despite the strengths of qualitative methods, the factors contributing to the successful adoption of biogas technology, as uncovered in this study, warrant further investigation. Mapping these factors to determine which have the greatest influence on biogas practices and usage in daily life is crucial. Given that energy consumption intersects with various issues such as cultural norms, social interactions, and religious practices these elements should be thoroughly examined in future research.

References

Bijker, W. E. (2001). Understanding technological culture through a constructivist view of science, technology and society. In S. H. Cutcliffe & C, Mitcham (Eds.), *Visions of STS: Counterpoints in science, technology and society studies* (pp. 19–34). State University of New York Press.

Bijker, W. E. (2009). Globalization and vulnerability: Challenges and opportunities for SHOT around its fiftieth anniversary. *Technology and Culture, 50*(3), 600–612. https://doi.org/10.1353/tech.0.0313

Cavicchi, B. (2016). Sustainability that backfires: The case of biogas in Emilia Romagna. *Environmental Innovation and Societal Transitions, 21*, 13–27. https://doi.org/10.1016/j.eist.2016.02.001

Chodkowska-Miszczuk, J. (2021). A new narrative for sustainability: Exploring biogas plants as "first movers" in raising energy awareness. *Australian Journal of Environmental Education, 38*(2), 1–16. https://doi.org/10.1017/aee.2021.17

Dobers, G. M. (2019). Acceptance of biogas plants taking into account space and place. *Energy Policy, 135*, 110987. https://doi.org/10.1016/j.enpol.2019.110987

Flichy, P. (2007). *Understanding technological innovation: A socio-technical approach*. Edward Elgar.

Forbord, M., & Hansen, L. (2020). Enacting sustainable transitions: A case of biogas production and public transport in Trøndelag, Norway. *Journal of Cleaner Production, 254*, 120156. https://doi.org/10.1016/j.jclepro.2020.120156

Hasan, A. S. M. M., Kabir, M. A., Hoq, M. T., Johansson, M. T., & Thollander, P. (2022). Drivers and barriers to the implementation of biogas technologies in Bangladesh. *Biofuels, 13*(5), 643–655. https://doi.org/10.1080/17597269.2020.1841362

Hewitt, J., Holden, M., Robinson, B. L., Jewitt, S., & Clifford, M. J. (2022). Not quite cooking on gas: Understanding biogas plant failure and abandonment in Northern Tanzania. *Renewable and Sustainable Energy Reviews, 165*, 112600. https://doi.org/10.1016/j.rser.2022.112600

Ismail, C., Wiropranoto, F., Takama, T., Lieu, J., & Virla, L. D. (2021). Frugal eco-innovation for addressing climate change in emerging countries: Case of biogas digester in Indonesia. In *Handbook of climate change management: Research, leadership, transformation* (pp. 693–719). Springer.

Jalas, M., Kuusi, H., & Heiskanen, E. (2014). Self-Building courses of solar heat collectors as sources of consumer empowerment and local embedding of sustainable energy technology. *Science & Technology Studies, 27*(1), 76–96. https://doi.org/10.23987/sts.55335

Kanda, W., & Hjelm, O. (2021). Drivers for and barriers to the diffusion of biogas technologies through export. *Technological Forecasting and Social Change, 168*. https://doi.org/10.1016/j.techfore.2021.120780

Lehtonen, T.-K. (2003). The domestication of new technologies as a set of trials. *Journal of Consumer Culture, 3*(3), 363–385. https://doi.org/10.1177/14695405030033014

Mackay, H., & Gillespie, G. (1992). Extending the social shaping of technology approach: Ideology and appropriation. *Social Studies of Science, 22*(4), 685–716. https://doi.org/10.1177/030631292022004006

Meng, Z., He, J., & Xu, D. (2023). How do peer effects affect the transformation of farmers' willingness and behavior to adopt biogas? *Journal of Cleaner Production, 415*, 137857–137857. https://doi.org/10.1016/j.jclepro.2023.137857

Selwyn, N. (2003). Apart from technology: Understanding people's non-use of information and communication technologies in everyday life. *Technology in Society, 25*(1), 99–116. https://doi.org/10.1016/s0160-791x(02)00062-3

Sime, G. (2020). Technical and socioeconomic constraints to the domestication and functionality of biogas technology in rural areas of southern Ethiopia. *Cogent Engineering, 7*(1), 1765686. https://doi.org/10.1080/23311916.2020.1765686

Sørensen, K. H. (2006). Domestication; the enactment of technology. In T. Berker, M. Hartmann, Y. Punie, & K. Ward (Eds.), *Domestication of media and technology*. Open University Press.

Tembo, A., Rahman, M. M., & Jerin, T. (2023). Barriers to development and adoption of biogas in Mokambo peri-urban of Mufulira, Zambia: How does local government fail to provide renewable energy? *Biofuels, 14*(6), 583–594. https://doi.org/10.1080/17597269.2022.2156055

Verbeek, P. P., & Vermaas, P. E. (2010). Technological artifacts. In *A companion to the philosophy of technology* (pp. 165–171). Blackwell.

Wajcman, J. (2004). *Techno feminism*. Polity Press.

Wang, Z., Ali, S., Akbar, A., & Rasool, F. (2020). Determining the influencing factors of biogas technology adoption intention in Pakistan: The moderating role of social media. *International Journal of Environmental Research and Public Health, 17*(7), 2311. https://doi.org/10.3390/ijerph17072311

Ward, K. (2006). The bald guy just ate an orange. Domestication, work, and home. In M. Hartmann, Y. Punie, & K. Ward (Eds.), *Domestication of media and technology*. Open University Press.

The manufacturer's authorised representative in the EU is Springer Nature Customer Service Centre GmbH, Europaplatz 3, 69115 Heidelberg, Germany. If you have any concerns regarding our products, please contact ProductSafety@springernature.com

Printed and bound by CPI Group (UK) Ltd, Croydon, CR0 4YY
30/09/2026
02225309-0012